SpringerBriefs in Speech Technology

Series Editor:
Amy Neustein

For further volumes:
http://www.springer.com/series/10043

Editor's Note

The authors of this series have been hand-selected. They comprise some of the most outstanding scientists –drawn from academia and private industry –whose research is marked by its novelty, applicability, and practicality in providing broad based speech solutions. The SpringerBriefs in Speech Technology series provides the latest findings in speech technology gleaned from comprehensive literature reviews and *empirical investigations* that are performed in both laboratory and *real life* settings. Some of the topics covered in this series include the presentation of real life commercial deployment of spoken dialog systems, contemporary methods of speech parameterization, developments in information security for automated speech, forensic speaker recognition, use of sophisticated speech analytics in call centers, and an exploration of new methods of soft computing for improving human-computer interaction. Those in academia, the private sector, the self service industry, law enforcement, and government intelligence, are among the principal audience for this series, which is designed to serve as an important and essential reference guide for speech developers, system designers, speech engineers, linguists and others. In particular, a major audience of readers will consist of researchers and technical experts in the automated call center industry where speech processing is a key component to the functioning of customer care contact centers.

Amy Neustein, Ph.D., serves as Editor-in-Chief of the International Journal of Speech Technology (Springer). She edited the recently published book "Advances in Speech Recognition: Mobile Environments, Call Centers and Clinics" (Springer 2010), and serves as quest columnist on speech processing for Womensenews. Dr. Neustein is Founder and CEO of Linguistic Technology Systems, a NJ-based think tank for intelligent design of advanced natural language based emotion-detection software to improve human response in monitoring recorded conversations of terror suspects and helpline calls. Dr. Neustein's work appears in the peer review literature and in industry and mass media publications. Her academic books, which cover a range of political, social and legal topics, have been cited in the Chronicles of Higher Education, and have won her a pro Humanitate Literary Award. She serves on the visiting faculty of the National Judicial College and as a plenary speaker at conferences in artificial intelligence and computing. Dr. Neustein is a member of MIR (machine intelligence research) Labs, which does advanced work in computer technology to assist underdeveloped countries in improving their ability to cope with famine, disease/illness, and political and social affliction. She is a founding member of the New York City Speech Processing Consortium, a newly formed group of NY-based companies, publishing houses, and researchers dedicated to advancing speech technology research and development.

Fathi E. Abd El-Samie

Information Security for Automatic Speaker Identification

 Springer

Fathi E. Abd El-Samie
Department of Electronics and Electrical Communications
Faculty of Electronic Engineering
Menoufia University
Menouf, 32952, Egypt
Fathi_sayed@yahoo.com

ISSN 2191-737X   e-ISSN 2191-7388
ISBN 978-1-4419-9697-8   e-ISBN 978-1-4419-9698-5
DOI 10.1007/978-1-4419-9698-5
Springer New York Dordrecht Heidelberg London

Library of Congress Control Number: 2011929038

Printed on acid-free paper

Springer is part of Springer Science+Business Media (www.springer.com)

To my grand mother, my mother, my wife, and my daughter.

Preface

I know what you are asking yourself – "there are a lot of books available in speech processing, what is novel in this book?" Well, I can summarize the answer for this question in the following points:

1. You always see different algorithms for speech enhancement, deconvolution, signal separation, watermarking, and encryption, separately, without specific applications for these algorithms.
2. You also see literature books and research papers on speaker identification concentrating on how to extract features, and the comparison between feature extraction methods.
3. How to make use of speech enhancement, deconvolution, and signal separation to enhance the performance of speaker identification systems is a missing issue. This book presets this issue and gives comparison studies between different algorithms that can be used for this purpose.
4. Speech watermarking and encryption are studied for the first time in this book in a framework that enhances the security of speaker identification systems.
5. Performance enhancement and security enhancement of speaker identification systems are contradicting objectives. How they affect each other is also studied in this book.

Finally, I hope that this book will be a starting step towards an extensive study to build speaker identification systems with multilevels of security.

Fathi E. Abd El-Samie

Acknowledgements

I would like to thank my students Amira Shafik, Hossam Hammam, Emad Mosa, and Marwa Abdelfattah for their cooperation in software arrangement.

Contents

About the Author

Fathi E. Abd El-Samie received the B.Sc. (Honors), M.Sc., and Ph.D. from the Faculty of Electronic Engineering, Menoufia University, Menouf, Egypt, in 1998, 2001, and 2005, respectively. He joined the teaching staff of the Department of Electronics and Electrical Communications, Faculty of Electronic Engineering, Menoufia University, Menouf, Egypt, in 2005. He is a coauthor of about 130 papers in international conference proceedings and journals. He has received the most cited paper award from Digital Signal Processing journal for 2008. His current research areas of interest include image processing, speech processing, and digital communications.

Information Security for Automatic Speaker Identification

Abstract Speaker identification is a widely used technique in several security systems. In remote access systems, speaker utterances are recoded and communicated through a communication channel to a receiver that performs the identification process. Speaker identification is based on characterizing each speaker with a set of features extracted from his or her utterance. Extracting the features from a clean speech signal guarantees the high success rate in the identification process. In real cases, a clean speech is not available for feature extraction due to channel degradations, background noise, or interfering audio signals. As a result, there is a need for speech enhancement, deconvolution, and separation algorithms to solve the problem of speaker identification in the presence of impairments. Another important issue, which deserves consideration, is how to enhance the security of a speaker identification system. This can be accomplished by watermark embedding in the clean speech signals at the transmitter. If this watermark is extracted correctly at the receiver, it can be used to ensure the correct speaker identification. Another means of security enhancement is the encryption of speech at the transmitter. Speech encryption prevents eavesdroppers from getting the speech signals that will be used for feature extraction to avoid any unauthorized access to the system by synthesis trials. Multilevels of security can be achieved by implementing both watermarking and encryption at the transmitter. The watermarking and encryption algorithms need to be robust to speech enhancement, and deconvolution algorithms to achieve the required degree of security and the highest possible speaker identification rates. This book provides for the first time a comprehensive literature review on how to improve the performance of speaker identification systems in noisy environments, by combining different feature extraction techniques with speech enhancement, deconvolution, separation, watermarking, and/or encryption.

Keywords Speech enhancement • Speech deconvolution • Signal separation • Speech watermarking • Speech encryption • Wavelet denoising • Wiener filter • Singular value decomposition • Chaotic baker map

F.E.A. El-Samie, *Information Security for Automatic Speaker Identification*,
SpringerBriefs in Speech Technology, DOI 10.1007/978-1-4419-9698-5_1,
© Springer Science+Business Media, LLC 2011

1   Introduction

Automatic speaker identification involves recognizing a person from his spoken words [1–3]. The goal of speaker identification is to find a unique voice signature to discriminate one person from another. The techniques involved with this task can be classified into identification and verification techniques. Speaker identification is the process of determining which registered speaker provides a given utterance. Speaker verification is the process of accepting or rejecting the identity claim of a speaker.

The speaker identification process may be text dependent or text independent. In text dependent speaker identification systems, the speaker is asked to utter a specific string of words both in the training and recognition phases, whereas in text independent systems, the speaker identification system recognizes the speaker irrespective of any specific phrase utterance. Speaker identification systems can be open set or closed set. In closed set systems, the speaker is known a priori to be a member of a set of finite speakers. In open set systems, there is also an additional possibility of a speaker being an outsider i.e., not from the set of already defined speakers.

Speaker identification systems have several applications such as voice dialing, banking by telephone, telephone shopping, database access services, information services, voice mail, security control for confidential information areas, remote access to computers, controlling access to computer networks and websites, law enforcement, prison call monitoring, and forensic analysis [2, 4]. These systems contain two main processes; feature extraction and classification. Feature extraction extracts a small amount of data from the speech signal that can be used later to represent each speaker. There are various techniques for extracting speech features such as the Mel-frequency cepstral coefficients (MFCCs) technique. This technique is widely used in several applications such as speaker identification [1–3], fingerprint identification [5], landmine detection [6, 7], defect detection in industrial applications [8], and device modeling [9–12] due to its ability to characterize a large amount of data with a few features. Classification is a process having two phases; speaker modeling and speaker matching. Classification in this book is based on artificial neural networks (ANNs).

The MFCCs are not robust enough in noisy environments. This problem is solved by extracting MFCCs from transform domains rather than the time domain [13]. Transforms such as the discrete cosine transform (DCT) and the discrete sine transform (DST) enjoy a sophisticated energy compaction property, which can be efficiently utilized for feature extraction. Another popular transform; the discrete wavelet transform (DWT), decomposes the signal into subbands leading to distinguishing features for each subband.

Speaker identification in the presence of noise, interference, or channel degradations is a challenging task. Speech enhancement, separation, and deconvolution techniques can be utilized to enhance the performance of speaker identification systems in the presence of degradations. Techniques like the spectral subtraction, Wiener filtering, adaptive Wiener filtering, and wavelet denoising can be used to enhance speech signals before the feature extraction process. For channel degradations, deconvolution techniques such as the linear minimum mean square (LMMSE) and regularized deconvolution are useful.

As mentioned above, the objective of speaker identification systems is to get a voice signature for each speaker to achieve a high degree of security in certain applications. The main objective of this book is to provide a novel speaker identification system with multilevels of security based speech watermarking, speech encryption, and speech signal processing. Multilevels of security can be achieved in speaker identification systems by incorporating speech watermarking and speech encryption with these systems. In some applications of speaker identification, speech signals are communicated through a channel prior to the identification process. It is possible for a watermarking process, an encryption process, or both of them to be performed at the transmitter. At the receiver, a watermark extraction, a decryption, or both of them can be performed. For the case of watermarking, an image for example can be embedded in the speech signal at the transmitter. If this watermark is extracted successfully and the extracted features form the speech signal match a candidate's features in the database, this can be used as a double check for an authorized speaker. Encryption can also be used at the sender to hide the identity of the speaker from an eavesdropper in the channel, who can alter the speaker voice features, or synthesize another speech signal once again for the speaker.

2   Speaker Identification

In speaker identification, a speech utterance from an unknown speaker is analyzed and compared with models of all known speakers. The unknown speaker is identified as the speaker, whose model best matches the input utterance. Speaker identification involves three stages; feature extraction to represent the speaker information present in the speech signal, modeling of the speaker's features, and decision making to complete the identification task. The main task in a speaker identification system is to extract features capable of representing the speaker information present in the speech signal. Once a proper set of feature vectors is obtained, the next task is to develop a model for each speaker. Feature vectors representing the voice characteristics of the speaker are extracted and used for building the reference models. The final stage is the decision to either accept or reject the claim of the speaker. This decision is made based on the result of the matching technique used. The block diagram of a speaker identification system is shown in Fig. 1.

The speaker identification process consists of two modes; a training mode and recognition or testing mode as shown in Fig. 2 [14]. In the training mode, a new speaker with known identity is enrolled into the system database. In the recognition mode, an unknown speaker gives a speech input and the system makes a decision about the speaker's identity.

Both the training and the recognition modes include a feature extraction step, which converts the digital speech signal into a sequence of numerical features, called feature vectors. The feature vectors provide a more stable, robust, and compact representation than the raw input speech signal. Feature extraction can be considered as a data reduction process that attempts to preserve the essential characteristics of the speaker, while removing any redundancy. Features are

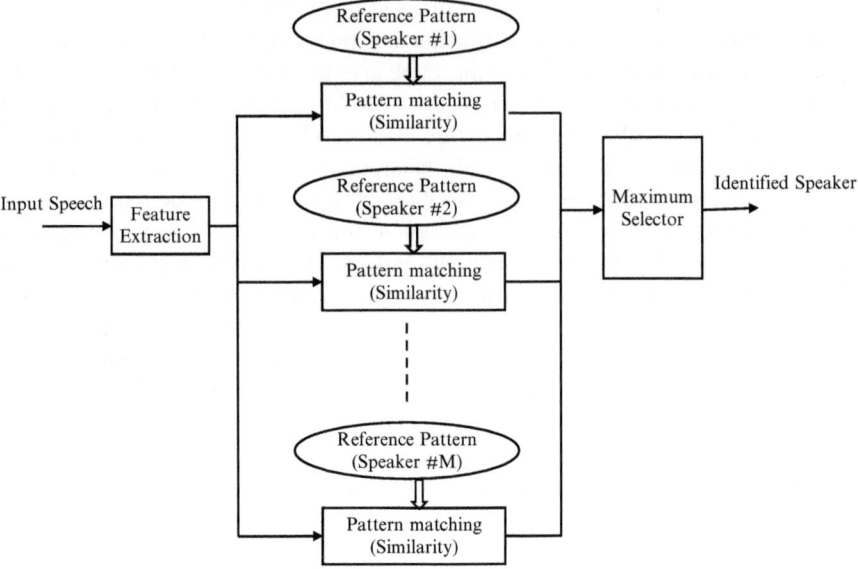

Fig. 1 Block diagram of the speaker identification system

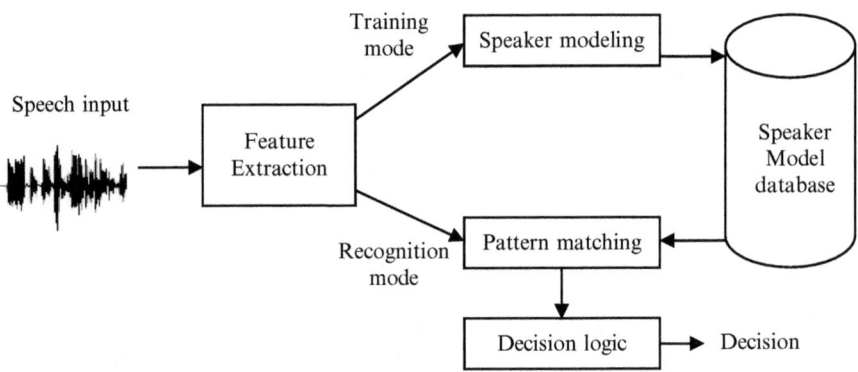

Fig. 2 Training and testing modes of an automatic speaker identification system

extracted from the training data essentially filtering out all unnecessary information in the training speech samples leaving only the speaker's characteristic information, with which the speaker's model can be constructed. In the recognition mode, features are extracted from the unknown speaker's voice.

Pattern matching refers to the algorithm that computes a matching score between the unknown speaker's feature vector and the models stored in the database. The output of the pattern matching module is a similarity score. The last phase in the recognition system is the decision making. The decision-making module takes the matching score as its input, and makes the final decision of the speaker's identity.

2.1 Feature Extraction

The methodology of the human brain to distinguish between speakers is based on high-level features such as dialect, speaking style, and emotional state. Although these features can characterize the speaker efficiently, it is difficult to build a speaker identification system based on them due to the large complexity problem. So, the alternative is to build the speaker identification system based on low-level features. An example of such low-level features is the MFCCs.

To extract the MFCCs from a speech signal, it is necessary to investigate how this signal is generated (Fig. 3). A speech signal $s(n)$ can be expressed in terms of an excitation $e(n)$ and a vocal tract model $h(n)$ as a convolution in the time domain [15]:

$$s(n) = h(n) * e(n), \tag{1}$$

where $e(n)$ is the excitation and $h(n)$ is the vocal tract impulse response.

The idea of cepstral analysis is to separate the spectral components of the excitation and the vocal tract, so that speech or speaker dependent information represented by the vocal tract can be obtained. Mathematically, the cestrum is computed by taking the fast Fourier transform (FFT) of the signal, the log of the magnitude spectrum, and then the inverse fast Fourier transform (IFFT) as follows:

$$\text{Cepstrum(frame)} = \text{FFT}^{-1}(\log(|\text{FFT(frame)}|)). \tag{2}$$

In the time domain, a convolution relationship exists as shown in (Eq. 1). Taking the FFT moves the analysis to the frequency domain giving:

$$S(k) = H(k)E(k). \tag{3}$$

Taking the logarithm of (Eq. 3), the multiplied spectra become additive as follows:

$$\text{Log}|S(k)| = \text{Log}|H(k)| + \text{Log}|E(k)|. \tag{4}$$

The IFFT can then be taken. It operates on the two parts of (Eq. 4), separately, resulting in the cepstral representation of the signal. It is possible to separate the excitation spectrum $E(k)$ from the vocal tract system spectrum $H(k)$ taking into account the fact that $E(k)$ is responsible for the fast spectral variations, and $H(k)$ is responsible for the slow spectral variations. The domain created after taking the logarithm and the IFFT is called the cepstral domain, and the word quefrency is used for describing the frequencies in the cepstral domain.

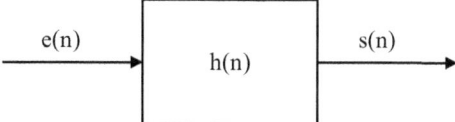

Fig. 3 Simple model of speech production

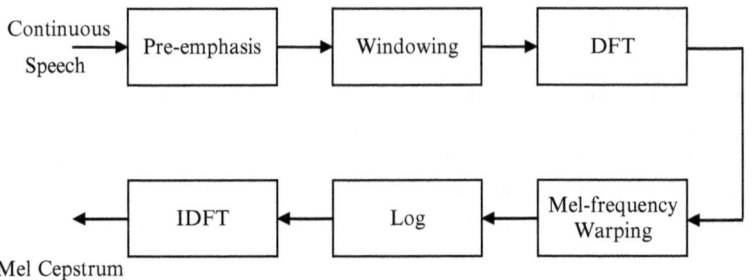

Fig. 4 Extraction of MFCCs from a speech signal

In the cepstral domain, the excitation signal and the vocal tract impulse response can be separated using a lifter. The vocal tract response decides the spectral envelope, and exists in the low quefrency region, while the excitation (pitch information) represents the spectral details, and exists in the high quefrency region. For speaker identification, the spectral envelope is more useful than the spectral details.

For the calculation of the MFCCs of a speech signal, the signal is first framed and windowed, the DFT is then taken, and the magnitude of the resulting spectrum is warped by the Mel scale. The log of this spectrum is then taken and the DCT is applied. This is illustrated in Fig. 4. The steps of extraction of the MFCCs are summarized in the following subsections.

2.1.1 Preemphasis

The digitalized speech is preemphasized with a first-order finite impulse response (FIR) filter, for its linear phase and simple implementation. Since in speech signals, the lower formants often contain more energy, and therefore are preferentially modeled with respect to the higher formants, a preemphasis filter is therefore used to boost the high frequencies [16–18]. The digitalized speech is preemphasized to remove glottal and lip radiation effects. The preemphasis filter transfer function is given by:

$$H(z) = 1 - az^{-1}, \tag{5}$$

where $0.9 \le a \le 0.99$.

2.1.2 Framing and Windowing

The speech signal is a slowly time-varying signal. In a speaker identification system, the speech signal is partitioned into short-time segments called frames. To make the frame parameters vary smoothly, there is normally a 50% overlap between each two adjacent frames. Windowing is performed on each frame with one of the popular signal processing windows like the Hamming window [19]. Windowing is often applied to increase the continuity between adjacent frames and smooth out the end points such that abrupt changes between ends of successive frames are minimized.

As a frame is multiplied by a window, most of the data at the edge of the frame becomes insignificant causing loss of information. An approach to tackle this problem is to allow overlapping in the sections between frames, which allows adjacent frames to include portions of data in the current frame. This means that the edges of the current frame are included as the center data of adjacent frames. Typically, around 50% of overlapping is sufficient to embrace the lost information.

2.1.3   The DFT

Fourier analysis provides a way of analyzing the spectral properties of a given signal in the frequency domain. The Fourier transform converts a discrete signal $s(n)$ from time domain into frequency domain with the equation [19]:

$$S(k) = \sum_{n=0}^{N-1} s(n) e^{-j2\pi nk/N}, \quad 0 \leq k \leq N - 1, \tag{6}$$

where $n = 0, 1, \ldots, N - 1$, and N is the number of samples in the signal $s(n)$. k represents the discrete frequency index and j is equal to $\sqrt{-1}$. The result of the DFT is a complex-valued sequence of length N.

The IDFT is defined as:

$$s(n) = \frac{1}{N} \sum_{k=0}^{N-1} S(k) e^{j2\pi nk/N}, \quad 0 \leq n \leq N - 1. \tag{7}$$

2.1.4   The Mel Filter Bank

Psychophysical studies have shown that human perception of the frequency contents of sounds for speech signals does not follow a linear scale. In the MFCCs method, the main advantage is that it uses Mel-frequency scaling, which approximates quite well the human auditory system. The Mel scale is defined as [19]:

$$\text{Mel}(f) = 2,595 \ \log\left(1 + \frac{f}{700}\right), \tag{8}$$

where Mel is the Mel-frequency scale and f is the frequency on the linear frequency scale.

The MFCCs are extracted using a Mel filter bank, where the filters are spaced on the Mel scale approximately linearly below 1 kHz, and logarithmically above 1 kHz. The conventional Mel filter bank in speaker identification is composed of a number of triangular bandpass filters distributed inside the signal bandwidth.

2.1.5   The DCT

The final stage involves performing a DCT on the log of the Mel spectrum. If the output of the mth Mel filter is $\tilde{S}(m)$, then the MFCCs are given as [20]:

$$c_g = \sqrt{\frac{2}{N}} \sum_{m=1}^{N_f} \log(\tilde{S}(m)) \, \cos\left(\frac{g\pi}{N_f}(m - 0.5)\right), \tag{9}$$

where $g = 0, 1, \ldots, G - 1$, G is the number MFCCs, N_f is the number of Mel filters and c_g is the gth MFCC. The number of the resulting MFCCs is chosen between 12 and 20, since most of the signal information is represented by the first few coefficients. The 0th coefficient represents the mean value of the input signal.

2.1.6   Polynomial Coefficients

MFCCs are sensitive to channel mismatches between training and testing data, and they are also speaker dependent. Polynomial coefficients are added to the MFCCs to solve this problem. They help in increasing the similarity between the training and testing utterances, if they are related to the same person [21]. The importance of these coefficients arises from the fact that they can preserve valuable information (mean, slope, and curvature) about the shapes of the time function of each cepstral coefficient of the training and testing utterances.

When the person says the same word at two different times (training and testing), the amplitudes of a particular cepstral coefficient through frames of the training utterance may differ from those of the testing utterance, which would lead to an increase in the distance between the utterances and a decrease in the efficiency of the matching process. On the other hand, the shape of time functions of both cepstral coefficients is the same or very similar. Consequently, both of them have the same values of polynomial coefficients, which is very helpful in the matching process [21].

To calculate the polynomial coefficients, the time waveforms of the cepstral coefficients are expanded by orthogonal polynomials. The following two orthogonal polynomials can be used [21]:

$$P_1(i) = i - 5, \tag{10}$$

$$P_2(i) = i^2 - 10i + 55/3. \tag{11}$$

To model the shape of the MFCCs time functions, a nine elements window at each MFCC is used. Based on this windowing assumption, the polynomial coefficients can be calculated as follows [21]:

$$a_g(t) = \frac{\sum_{i=1}^{9} P_1(i) c_g(t + i + 1)}{\sum_{i=1}^{9} P_1^2(i)}, \tag{12}$$

$$b_g(t) = \frac{\sum\limits_{i=1}^{9} P_2(i)c_g(t+i+1)}{\sum\limits_{i=1}^{9} P_2^2(i)}, \tag{13}$$

where $a_g(t)$ and $b_g(t)$ are the slope, and the curvature of the MFCCs time functions at each c_g. The vectors containing all c_g, a_g, and b_g are concatenated to form a single feature vector for each speech signal.

2.2   Feature Matching

The classification step in automatic speaker identification systems is in fact a feature matching process between the features of a new speaker and the features saved in the database. Neural Networks are widely used for feature matching. Multilayer perceptrons (MLPs) consisting of an input layer, one or more hidden layers, and an output layer can be used for this purpose [22, 23]. Figure 5 shows an MLP having an input layer, a single hidden layer, and an output layer. A single neuron only of the output layer is shown for simplicity. This structure can be used for feature matching in the speaker identification process.

Each neuron in the neural network is characterized by an activation function and its bias, and each connection between two neurons by a weight factor. In this paper, the neurons from the input and output layers have linear activation functions and

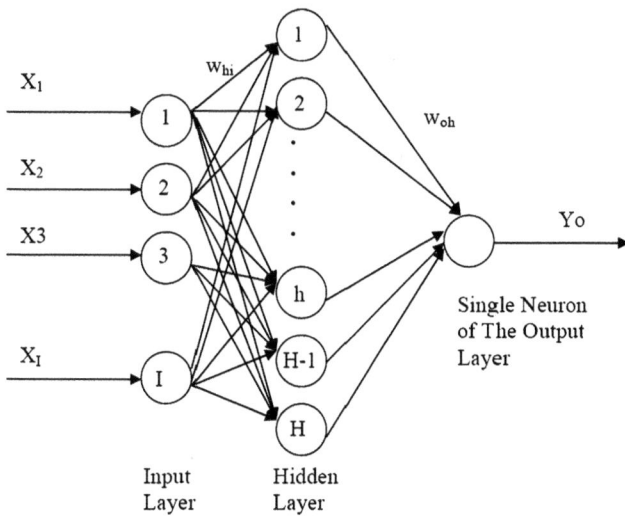

Fig. 5 An MLP neutral network

hidden neurons have sigmoid activation functions $F(u) = 1/(1 + e^{-u})$. Therefore, for an input vector $\mathbf{X}$, the neural network output vector $\mathbf{Y}$ can be obtained according to the following matrix equation [22, 23]:

$$\mathbf{Y} = \mathbf{W}_2 * F(\mathbf{W}_1 * \mathbf{X} + \mathbf{B}_1) + \mathbf{B}_2, \tag{14}$$

where $\mathbf{W}_1$ and $\mathbf{W}_2$ are the weight matrices between the input and the hidden layers, and between the hidden and the output layers, respectively. $\mathbf{B}_1$ and $\mathbf{B}_2$ are bias matrices for the hidden and the output layers, respectively.

Training a neural network is accomplished by adjusting its weights using a training algorithm. The training algorithm adapts the weights by attempting to minimize the sum of the squared error between a desired output and the actual output of the output neurons given by [22, 23]:

$$E = \frac{1}{2} \sum_{o=1}^{O} (D_o - Y_o)^2, \tag{15}$$

where D_o and Y_o are the desired and actual outputs of the oth output neuron. O is the number of output neurons. Each weight in the neural network is adjusted by adding an increment to reduce E as rapidly as possible. The adjustment is carried out over several training iterations until a satisfactorily small value of E is obtained or a given number of epochs is reached. The error back-propagation algorithm can be used for this task [22, 23].

3   Feature Extraction from Discrete Transforms

Discrete transforms can be used for extraction of robust MFCCs in speaker identification systems. The DWT, the DCT, and the DST have been investigated in the literature for this purpose [13]. Figure 6 illustrates the utilization of discrete transforms in speaker identification systems.

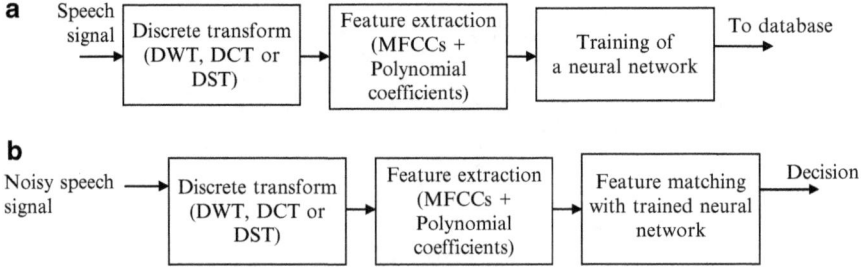

Fig. 6 Speaker identification based on discrete transforms. (**a**) Training phase. (**b**) Testing phase

3.1 The Discrete Wavelet Transform

It is known that the DFT considers the analysis of a speech signal separately in the time and frequency domains and does not provide temporal information about frequencies. Although the DFT may be a good tool for analyzing a stationary signal, speech signals are nonstationary or partially stationary. When analyzing a nonstationary signal, in addition to the frequency content of the signal, we need to know how the frequency content of the signal changes with time.

To overcome this deficiency, a modified transform called the short-time Fourier transform (STFT) has been adopted, because it allows the representation of the signal in both time and frequency domains through time widowing functions. The window length determines a constant time and frequency resolution. The main idea behind the STFT is to have localization in time domain. A drawback of the STFT is its small and fixed window, so that the STFT cannot capture the rapid changes in the signal. Moreover, it does not give information about the slowly changing parts of the signal [24].

Wavelet analysis provides an exciting alternative method to Fourier analysis for speech processing. Wavelet transform allows a variable time-frequency resolution, which leads to locality in both the time and frequency domains. The locality of the transform of a signal is important in two ways for pattern recognition. First, different parts of the signal may convey different amounts of information. Second, when the signal is corrupted by local noise in time and/or frequency domain, the noise affects only a few coefficients if the coefficients represent local information in the time and frequency domains.

In fact, the wavelet transform is a mathematical operation used to divide a given speech signal into different subbands of different scales to study each scale, separately. The idea of the DWT is to represent a signal as a series of approximation (lowpass version) and details (highpass version) at different resolutions. The speech signal is lowpass filtered to give an approximation signal and highpass filtered to give a detail signal. Both of them can be used to model the speech signal. The wavelet decomposition and reconstruction process is illustrated in Fig. 7.

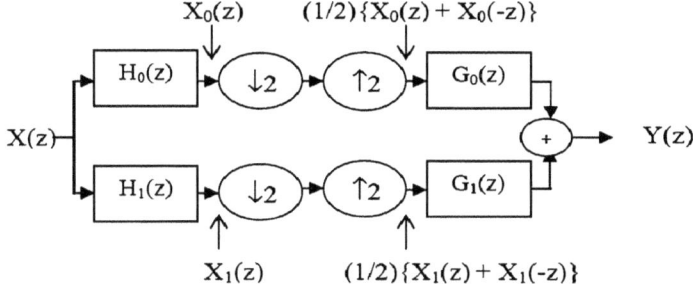

Fig. 7 The two band decomposition-reconstruction wavelet filter bank

The multilevel DWT can be regarded as equivalent to filtering the speech signal with a bank of bandpass filters, whose impulse responses are all approximately given by scaled versions of a mother wavelet. The scaling factor between adjacent filters is usually 2:1 leading to octave bandwidths and center frequencies that are one octave apart [24–38]. The outputs of the filters are usually maximally decimated so that the number of DWT output samples equals the number of input samples, and thus no redundancy occurs in this transform.

The art of finding a good wavelet lies in the design of the set of filters, H_0, H_1, G_0, and G_1 to achieve various tradeoffs between spatial and frequency domain characteristics, while satisfying the perfect reconstruction (PR) condition [35]. In Fig. 7, the process of decimation and interpolation by 2 at the outputs of H_0 and H_1 effectively sets all odd samples of these signals to zero. For the lowpass branch, this is equivalent to multiplying $x_0(n)$ by $1/2(1 + (-1)^n)$. Hence, $X_0(z)$ is converted to $1/2\{X_0(z) + X_0(-z)\}$. Similarly, $X_1(z)$ is converted to $1/2\{X_1(z) + X_1(-z)\}$.

As a result, the expression for $Y(z)$ is given by [35]:

$$Y(z) = \frac{1}{2}\{X_0(z) + X_0(-z)\}G_0(z) + \frac{1}{2}\{X_1(z) + X_1(-z)\}G_1(z)$$

$$= \frac{1}{2}X(z)\{H_0(z)G_0(z) + H_1(z)G_1(z)\} + \frac{1}{2}X(-z)\{H_0(-z)G_0(z) + H_1(-z)G_1(z)\}.$$

$$(16)$$

The first PR condition requires aliasing cancelation and forces the above term in $X(-z)$ to be zero. Hence, $\{H_0(-z)G_0(z) + H_1(-z)G_1(z)\} = 0$, which can be achieved if [35]:

$$H_1(z) = z^{-\rho}G_0(-z) \quad \text{and} \quad G_1(z) = z^{\rho}H_0(-z), \tag{17}$$

where ρ must be odd (usually $\rho = \pm 1$).

The second PR condition is that the transfer function from $X(z)$ to $Y(z)$ should be unity:

$$\{H_0(z)G_0(z) + H_1(z)G_1(z)\} = 2. \tag{18}$$

If we define a product $P(z) = H_0(z)G_0(z)$ and substitute from (Eq. 17) into (Eq. 18), then the PR condition becomes [35]:

$$H_0(z)G_0(z) + H_1(z)G_1(z) = P(z) + P(-z) = 2. \tag{19}$$

This needs to be true for all z and, since the odd powers of z in $P(z)$ cancel with those in $P(-z)$, it requires that $p_0 = 1$ and $p_v = 0$ for all v even and nonzero. The polynomial $P(z)$ should be a zero-phase polynomial to minimize distortion. In general, $P(z)$ is of the following form [35]:

$$P(z) = \cdots + p_5 z^5 + p_3 z^3 + p_1 z + 1 + p_1 z^{-1} + p_3 z^{-3} + p_5 z^{-5} + \cdots. \tag{20}$$

The design method for the PR filters can be summarized in the following steps [35]:

- Choose $p_1, p_3, p_5, \ldots$ to give a zero-phase polynomial $P(z)$ with good characteristics.
- Factorize $P(z)$ into $H_0(z)$ and $G_0(z)$ with similar lowpass frequency response.
- Calculate $H_1(z)$ and $G_1(z)$ from $H_0(z)$ and $G_0(z)$.

To simplify this procedure, we can use the following relation:

$$P(z) = P_t(Z) = 1 + p_{t,1}Z + p_{t,3}Z^3 + p_{t,5}Z^5 + \cdots, \tag{21}$$

where

$$Z = \frac{1}{2}(z + z^{-1}). \tag{22}$$

The Haar wavelet is the simplest type of wavelets. In the discrete form, Haar wavelets are related to a mathematical operation called the Haar transform. The Haar transform serves as a prototype for all other wavelet transforms [35]. Like all wavelet transforms, the Haar transform decomposes a discrete signal into two sub-signals of half its length. One of them is a running average or trend; the other is a running difference or fluctuation. This uses the simplest possible $P_t(Z)$ with a single zero at $Z = -1$. It is represented as follows [35]:

$$P_t(Z) = 1 + Z \quad \text{and} \quad Z = \frac{1}{2}(z + z^{-1}). \tag{23}$$

Thus,

$$P(z) = \frac{1}{2}(z + 2 + z^{-1})$$
$$= \frac{1}{2}(z + 1)(1 + z^{-1}) = G_0(z)H_0(z). \tag{24}$$

We can find $H_0(z)$ and $G_0(z)$ as follows:

$$H_0(z) = \frac{1}{2}(1 + z^{-1}), \tag{25}$$

$$G_0(z) = (z + 1). \tag{26}$$

Using (Eq. 17) with $\rho = 1$:

$$G_1(z) = zH_0(-z) = \frac{1}{2}z(1 - z^{-1}) = \frac{1}{2}(z - 1),$$
$$H_1(z) = z^{-1}G_0(-z) = z^{-1}(-z + 1) = (z^{-1} - 1). \tag{27}$$

The two outputs of $H_0(z)$ and $H_1(z)$ are concatenated to form a single vector of the same length as the original speech signal. The features are extracted from this vector and used for speaker identification.

3.2   The DCT

The DCT is a 1D transform with an excellent energy compaction property. For a speech signal $x(n)$, the DCT is represented by [35]:

$$X(k) = \alpha(k) \sum_{n=0}^{N-1} x(n) \, \cos\left(\frac{\pi(2n+1)k}{2N}\right), \quad k = 0, 1, 2, \ldots, N-1, \qquad (28)$$

where

$$\alpha(0) = \sqrt{\frac{1}{N}}, \quad \alpha(k) = \sqrt{\frac{2}{N}}.$$

The inverse discrete cosine transform (IDCT) is given by:

$$x(n) = \sum_{k=0}^{N-1} \alpha(k) X(k) \, \cos\left(\frac{\pi(2n+1)}{2N}\right), \quad n = 0, 1, 2, \ldots, N-1. \qquad (29)$$

The features are extracted from $X(k)$ and used for speaker identification.

3.3   The DST

The DST is another triangular transform with common properties with the DCT. The mathematical representation of the DST is given by [35]:

$$X(k) = \sum_{n=0}^{N-1} x(n) \, \sin\left(\frac{\pi}{N+1}(n-1)(k+1)\right), \quad k = 0, \ldots, N-1. \qquad (30)$$

The features are extracted from $X(k)$ and used for speaker identification.

3.4   Speaker Identification with Discrete Transforms

In the training phase of the speaker identification system, a database is first composed for 15 speakers. To generate this database, each speaker repeats a certain sentence 10 times. Thus, 150 speech signals are used to generate MFCCs and polynomial coefficients to form the feature vectors of the database. These features are used to train a neural network. In the testing phase, each one of these speakers is asked to say the sentence again and his speech signal is then degraded. Similar features to that used in the training are extracted from those degraded speech signals and used for matching.

The features used in all experiments are 13 MFCCs and 26 polynomial coefficients forming a feature vector of 39 coefficients for each frame of the speech signal. Seven methods for extracting features are adopted in the experiment. In the first method, the MFCCs and the polynomial coefficients are extracted from the speech signals, only. In the second one, the features are extracted from the DWT of the speech signals. In the third method, the features are extracted from both the original speech signals and the DWT of these signals and concatenated together. In the fourth method, the features are extracted from the DCT of the speech signals. In the fifth method, the features are extracted from both the original speech signals and the DCT of these signals and concatenated together. In the sixth method, the features are extracted from the DST of the speech signals. In the seventh method, the features are extracted from both the original speech signals and the DST of these signals and concatenated together. The recognition rate is used as the performance evaluation metric in all experiments. It is defined as the ratio of the number of success identifications to the total number of identification trials.

For the speech signals contaminated by additive White Gaussian noise (AWGN), it is clear from Fig. 8 that the DCT and the DWT are good competitors for robust feature extraction in the presence of AWGN. Features extracted from the original

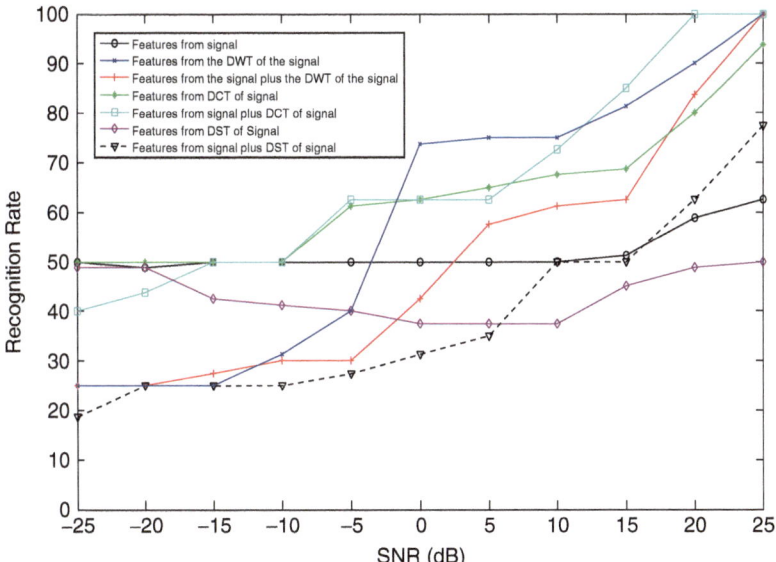

Fig. 8 Recognition rate vs. SNR for the different feature extraction methods for speech signals contaminated by AWGN

signals in the time domain fail to give good identification results, especially at low signal-to-noise ratios (SNRs). The DCT energy compaction property allows a few number features to characterize the speech signals, hence facilitating the matching process. Also, the subband decomposition resulting from the DWT allows feature extraction from the different bands to enhance the performance of speaker identification systems.

4 Speech Enhancement

Speech enhancement can be used as a preprocessing step in the testing phase of the speaker identification system to improve its performance as shown in Fig. 9.

4.1 Speech Quality Metrics

Speech quality metrics are used to assess the perceptual quality of the speech signals resulting from any speech enhancement algorithm. Several approaches, based on subjective and objective metrics, have been adopted in the literature for this purpose [39–42]. Objective metrics are generally divided into intrusive and nonintrusive. Intrusive metrics can be classified into three main groups. The first group includes time domain metrics such as the traditional SNR and the segmental signal-to-noise ratio (SNRseg). The second group includes linear predictive coefficients (LPCs) metrics, which are based on the LPCs of the speech signal and its derivative parameters, such as the linear reflection coefficients (LRCs), the log likelihood ratio (LLR), and the cepstral distance (CD). The third group includes the spectral domain metrics, which are based on the comparison between the power spectrum of the original signal and the processed signal. An example of such metrics is the spectral distortion (SD) [39–42].

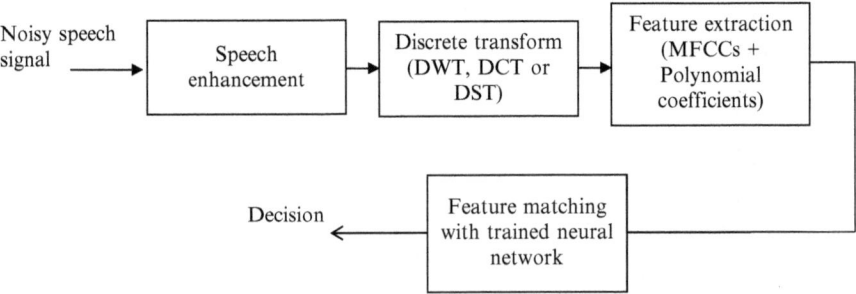

Fig. 9 Testing phase of a speaker identification system with speech enhancement

4.1.1   The SNR

The SNR is defined as follows [39–42]:

$$
\text{SNR} = 10 \log_{10} \frac{\sum\limits_{n=1}^{N} s^2(n)}{\sum\limits_{n=1}^{N} (s(n) - y(n))^2}, \tag{31}
$$

where $s(n)$ is the original speech signal, and $y(n)$ is the processed speech signal.

4.1.2   The SNRseg

SNRseg is defined as the average of the SNR values over short segments of the output signal. It is can be calculated as follows [39–42]:

$$
\text{SNRseg} = \frac{10}{M} \sum\limits_{m=0}^{M-1} \log_{10} \sum\limits_{n=L_s m}^{L_s m + L_s - 1} \left(\frac{s(n)}{(s(n) - y(n))} \right)^2, \tag{32}
$$

where M is the number of segments in the speech signal, and L_s is the length of each segment.

4.1.3   The LLR

The LLR metric for a speech segment is based on the assumption that the segment can be represented by an all-pole linear predictive coding model of the form [39–42]:

$$
s(n) = \sum\limits_{m=1}^{m_p} a_m s(n - m) + G_s u(n), \tag{33}
$$

where a_m (for $m = 1,2,\ldots,m_p$) are the coefficients of the all-pole filter, G_s is the gain of the filter, and $u(n)$ is an appropriate excitation source for the filter. The speech signal is windowed to form frames of 15–30 ms length. The LLR metric is then defined as [42]:

$$
\text{LLR} = \left| \log \left(\frac{\vec{a}_s \overline{R}_y \vec{a}_s^{\mathrm{T}}}{\vec{a}_y \overline{R}_y \vec{a}_y^{\mathrm{T}}} \right) \right|, \tag{34}
$$

where $\vec{a}_s$ is the LPCs coefficient vector $[1, a_s(1), a_s(2), \ldots, a_s(m_p)]$ for the original speech signal, $\vec{a}_y$ is the LPCs coefficient vector $[1, a_y(1), a_y(2), \ldots, a_y(m_p)]$ for the processed signal, and $\overline{\mathbf{R}}_y$ is the autocorrelation matrix of the processed speech signal. The closer the LLR to zero, the higher is the quality of the output speech signal.

4.1.4   The SD

The SD is a form of metrics that is implemented in frequency domain on the frequency spectra of the original and processed speech signals. It is calculated in dB to show how far is the spectrum of the processed signal from that of the original signal. The SD can be calculated as follows [39–42]:

$$SD = \frac{1}{M} \sum_{m=0}^{M-1} \sum_{k=L_s m}^{L_s m + L_s - 1} |V_s(k) - V_y(k)|, \tag{35}$$

where $V_s(k)$ is the spectrum of the original speech signal in dB for a certain segment and $V_y(k)$ is the spectrum of the processed speech signal in dB for the same segment. The smaller the SD, the better is the quality of the audio output signal.

4.2   Spectral Subtraction

The goal of the spectral subtraction method is the suppression of additive noise from the corrupted speech signal prior to speaker identification [43–45]. It is performed by subtracting the noise spectrum from the noisy signal spectrum to obtain an estimate of the clean signal spectrum, and then reconstructing the signal from the estimated spectrum. Speech degraded by additive noise can be represented by:

$$x(n) = s(n) + v(n), \tag{36}$$

where $s(n)$ is the clean speech signal, and $v(n)$ is the noise. Taking the Fourier transform gives:

$$X(k) = S(k) + V(k). \tag{37}$$

The spectral subtraction filter $H(k)$ is calculated by replacing the noise spectrum $V(k)$ with a spectrum, which can be readily measured. The magnitude $|V(k)|$ of $V(k)$ is replaced by its average value $\mu(k)$ taken during nonspeech activity, and the phase $\theta_V(k)$ of $V(k)$ is replaced by the phase $\theta_X(k)$ of $X(k)$. These substitutions result in the spectral subtraction estimated signal:

$$\hat{S}(k) = [|X(k)| - \mu(k)]e^{j\theta_X(k)} \tag{38}$$

or

$$\hat{S}(k) = H(k)X(k) \tag{39}$$

with

$$H(k) = 1 - \frac{\mu(k)}{|X(k)|} \tag{40}$$

and

$$\mu(k) = E\{|V(k)|\}. \tag{41}$$

The drawback of the spectral subtraction method is that it gives musical noise, which is an offensive noise. It is difficult to reduce the musical noise, because its spectrum is not stationary in short-time frames. Figure 10 shows a clean speech signal and its spectrogram. A contaminated version of this signal with an AWGN at an SNR = 5 dB is shown in Fig. 11. An enhanced version of the noisy signal using the spectral subtraction method is shown in Fig. 12. It is clear from that figure that effect of the spectral subtraction method is very slight. The effect of the spectral

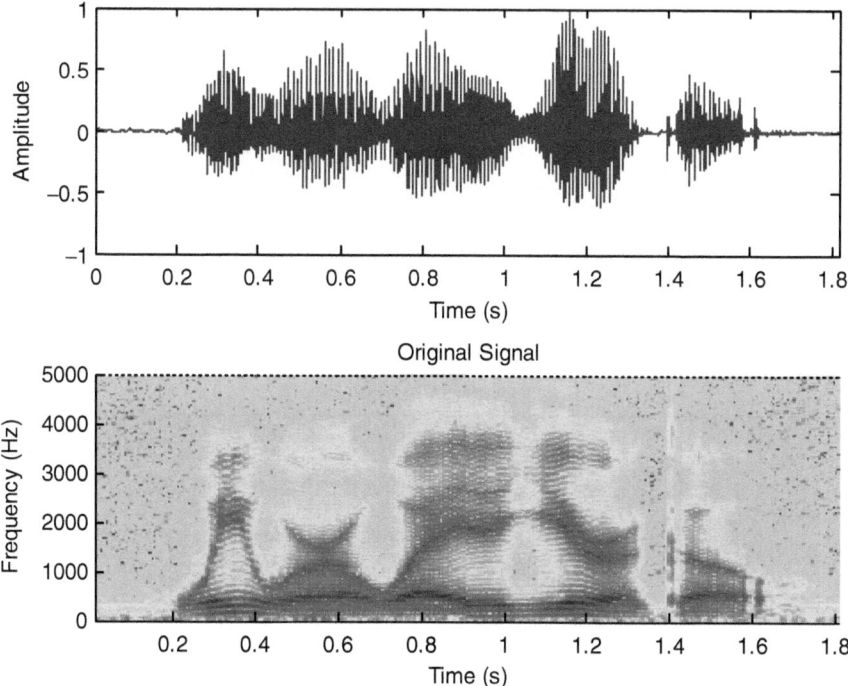

Fig. 10 Time domain waveform and spectrogram of a clean speech signal

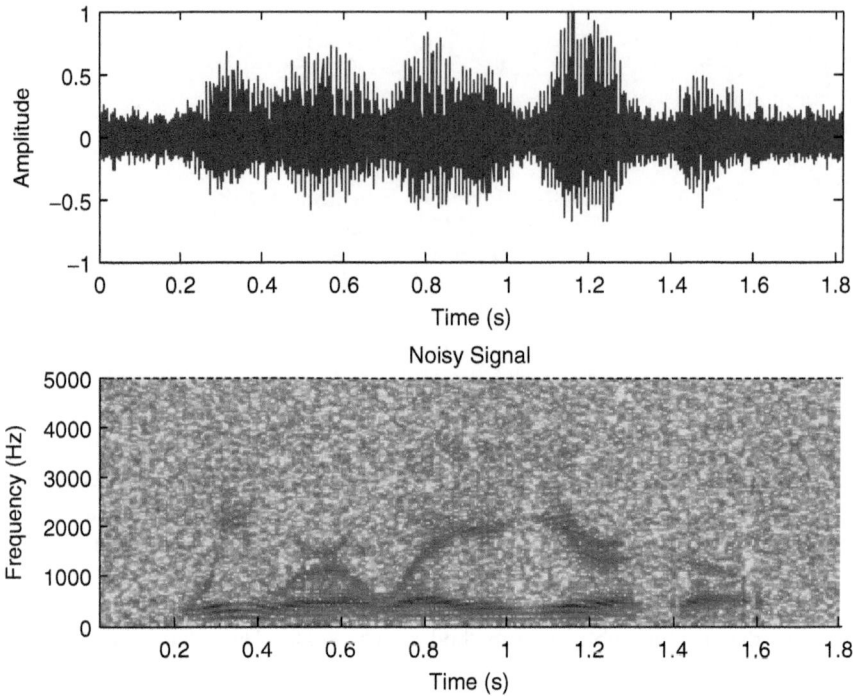

Fig. 11 Time domain waveform and spectrogram of the noisy signal with AWGN at SNR = 5 dB

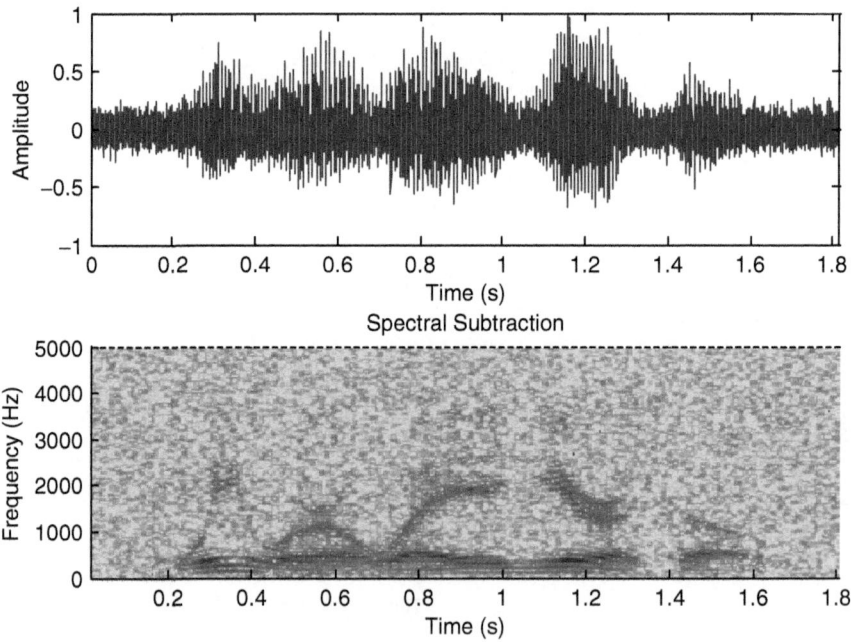

Fig. 12 Time domain waveform and spectrogram of the enhanced signal using the spectral subtraction method. SNR = 5.0439 dB, SNRseg = 5.0164 dB, LLR = 0.2336, SD = 8.4721 dB

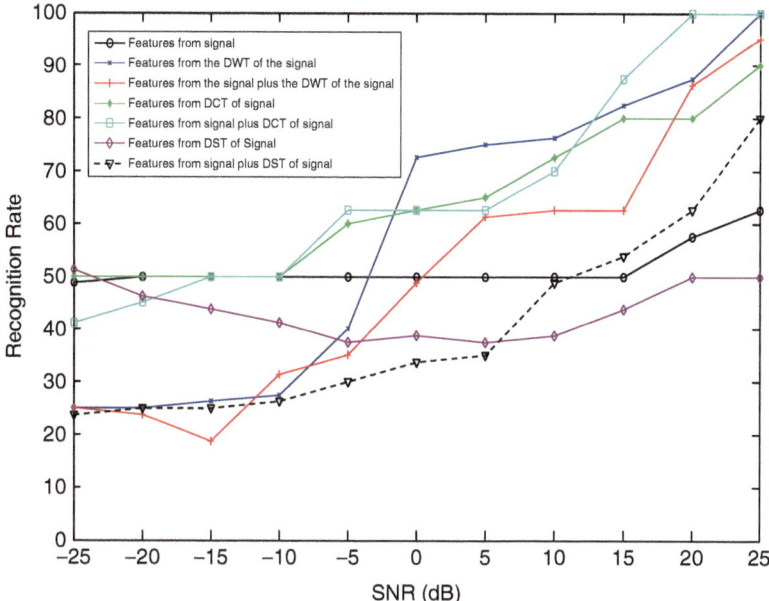

Fig. 13 Recognition rate vs. SNR for the different feature extraction methods in the presence of AWGN using the spectral subtraction method

subtraction method as a preprocessing method on the speaker identification process is shown in Fig. 13. It is clear from that figure that the effect of the spectral subtraction method on the process of speaker identification is also small.

4.3   Wiener Filter

The Wiener filter is an optimal filter that minimizes the mean square error (MSE) between the original and enhanced speech signals. This filter is defined by [46]:

$$S(k) = H(k)X(k), \qquad (42)$$

where $S(k)$, $X(k)$, and $H(k)$ are the DFT of the clean speech, the DFT of the noisy speech, and the transfer function of the Wiener filter, respectively. The Wiener filter is represented by:

$$H(k) = \frac{P_s(k)}{P_s(k) + P_v(k)}, \qquad (43)$$

where $P_s(k)$ and $P_v(k)$ are the power spectra of the speech signal $s(n)$ and the noise $v(n)$, respectively. This formula has been derived considering the signal $s(n)$ and noise $v(n)$ as uncorrelated and stationary signals. The SNR is defined by:

$$\text{SNR} = \frac{P_s(k)}{P_v(k)}. \tag{44}$$

This definition can be incorporated to Wiener filter equation as follows:

$$H(k) = \left[1 + \frac{1}{\text{SNR}}\right]^{-1}. \tag{45}$$

The drawback of the Wiener filter is the fixed frequency response at all frequencies and the requirement to estimate the power spectral density of the clean signal and noise prior to filtering. An enhanced version of the noisy speech signal using the Wiener filter is shown in Fig. 14. It is clear that the Wiener filtering method has a better performance than the spectral subtraction method. The effect of the Wiener filter enhancement on the speaker identification process is

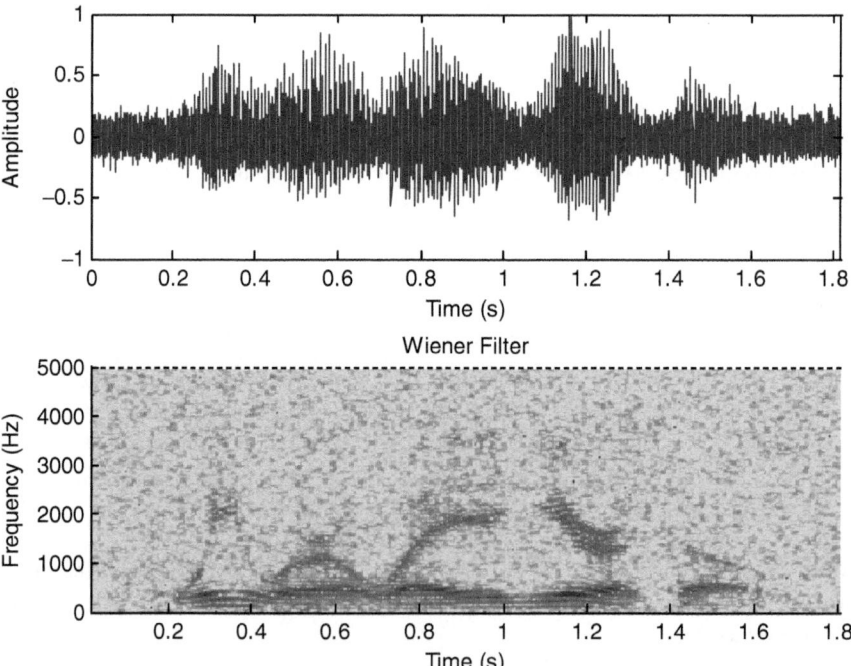

Fig. 14 Time domain waveform and spectrogram of the enhanced signal using the Wiener filtering method. SNR = 4.9880 dB, SNRseg = 4.9604 dB, LLR = 0.2383, SD = 8.5090 dB

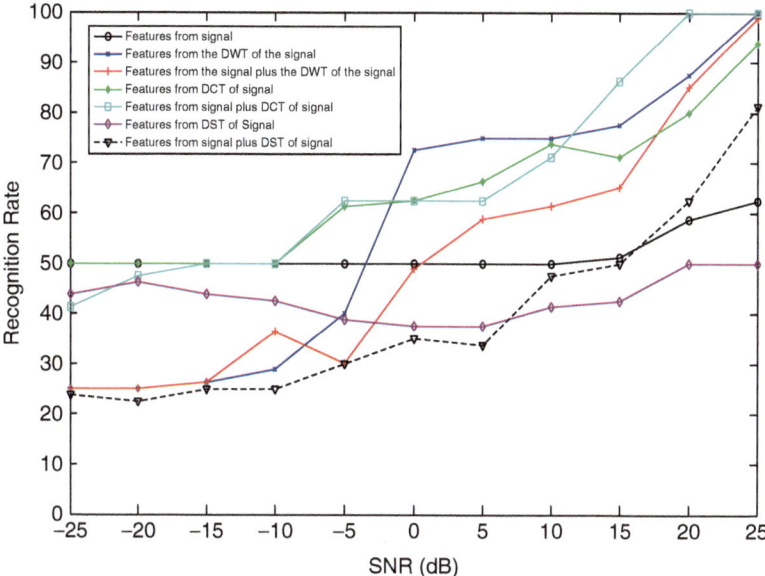

Fig. 15 Recognition rate vs. SNR for the different feature extraction methods in the presence of AWGN using the Wiener filter method

shown in Fig. 15. It is clear from that figure that the Wiener filtering method has a better effect on the speaker identification process than the spectral subtraction method.

4.4   The Adaptive Wiener Filter

The adaptive Wiener filter uses local statistics of the speech signal, and is derived from the Wiener filter under certain assumptions [46]. It is assumed that the additive noise $v(n)$ is a stationary white noise with zero mean and variance σ_v^2. Thus, the power spectrum of the noise $P_v(k)$ can be approximated by [46]:

$$P_v(k) = \sigma_v^2. \qquad (46)$$

Consider a small segment of the speech signal, in which the signal $x(n)$ is assumed to be stationary. The signal $x(n)$ can be modeled by [46]:

$$x(n) = m_x + \sigma_x w(n), \qquad (47)$$

where m_x and σ_x are the local mean and standard deviation of $x(n)$. $w(n)$ is a unit variance noise. Within this small segment of speech, the Wiener filter transfer function can be approximated by:

$$H(k) = \frac{P_s(k)}{P_s(k) + P_v(k)} = \frac{\sigma_s^2}{\sigma_s^2 + \sigma_v^2}. \tag{48}$$

Since $H(k)$ is constant over this small segment of speech, the impulse response of the Wiener filter can be obtained by:

$$h(n) = \frac{\sigma_s^2}{\sigma_s^2 + \sigma_v^2} \delta(n). \tag{49}$$

The enhanced speech signal $\hat{s}(n)$ in this local segment can be expressed as [46]:

$$\hat{s}(n) = m_x + (x(n) - m_x) * \frac{\sigma_s^2}{\sigma_s^2 + \sigma_v^2} \delta(n) = m_x + \frac{\sigma_s^2}{\sigma_s^2 + \sigma_v^2}(x(n) - m_x). \tag{50}$$

If m_x and σ_s are updated at each sample, we get:

$$\hat{s}(n) = m_x(n) + \frac{\sigma_s^2}{\sigma_s^2 + \sigma_v^2}(x(n) - m_x(n)). \tag{51}$$

In (Eq. 51), the local mean $m_x(n)$ and $(x(n) - m_x(n))$ are modified from segment to segment. If σ_s^2 is much larger than σ_v^2, the output signal $\hat{s}(n)$ will be primarily due to $x(n)$, and the input signal $x(n)$ is not attenuated. If σ_s^2 is smaller than σ_v^2, the filtering effect appears.

Note that m_x is identical to m_s when m_v is zero. So, we can estimate $m_x(n)$ in (Eq. 51) from $x(n)$ by:

$$\hat{m}_s(n) = \hat{m}_x(n) = \frac{1}{(2T + 1)} \sum_{l=n-T}^{n+T} x(l), \tag{52}$$

where $(2T + 1)$ is the number of samples in the short segment used in the estimation.

To measure the local statistics of the speech signal, we need to estimate the signal variance σ_s^2. Since $\sigma_x^2 = \sigma_s^2 + \sigma_v^2$, then $\sigma_s^2(n)$ may be estimated from $x(n)$ as follows:

$$\hat{\sigma}_x^2(n) = \begin{cases} \hat{\sigma}_x^2(n) - \hat{\sigma}_v^2 & \text{if } \hat{\sigma}_x^2(n) > \hat{\sigma}_v^2, \\ 0 & \text{otherwise,} \end{cases} \tag{53}$$

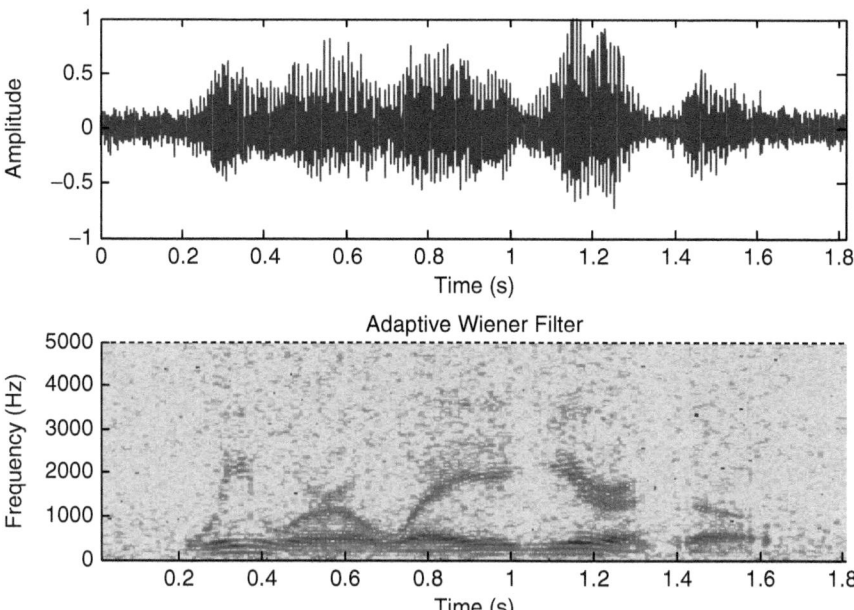

Fig. 16 Time domain waveform and spectrogram of the enhanced signal using the adaptive Wiener filtering method. SNR = 6.8726 dB, SNRseg = 6.8423 dB, LLR = 0.1609, SD = 7.3006 dB

where

$$\hat{\sigma}_x^2(n) = \frac{1}{(2T+1)} \sum_{l=n-T}^{n+T} (x(l) - \hat{m}_x(n))^2. \tag{54}$$

An enhanced version of the noisy speech signal using the adaptive Wiener filter is shown in Fig. 16. It is clear that the adaptive Wiener filtering method has a better performance than both the spectral subtraction method and the Wiener filtering method. The effect of the adaptive Wiener filtering on the speaker identification process is shown in Fig. 17. It is clear from that figure that the adaptive Wiener filter has a better effect on the speaker identification process than the spectral subtraction and the Wiener filtering methods.

4.5 Wavelet Denoising

Wavelet denoising is a simple operation, which aims at reducing noise in a noisy speech signal. It is performed by choosing a threshold that is sufficiently a large multiple of the standard deviation of the noise in the speech signal. Most of the

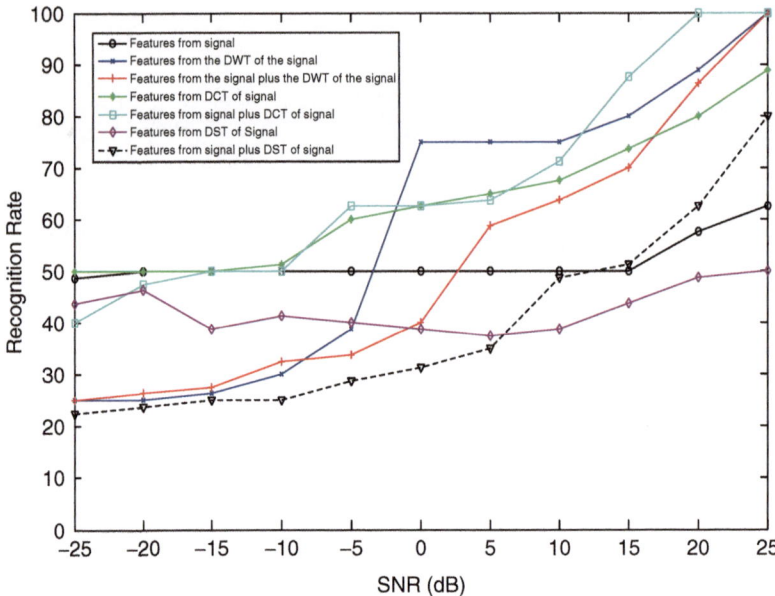

Fig. 17 Recognition rate vs. SNR for the different feature extraction methods in the presence of AWGN using the adaptive Wiener filtering method

noise power is removed by thresholding the detail coefficients of the wavelet transformed speech signal. There are two types of thresholding; hard and soft thresholding. The equation of the hard thresholding is given by [37, 47–49]:

$$f_{\text{hard}}(x_w) = \begin{cases} x_w, & |x_w| \geq \text{TH}, \\ 0, & |x_w| < \text{TH}. \end{cases} \tag{55}$$

On the other hand, that of soft thresholding is given by:

$$f_{\text{soft}}(x_w) = \begin{cases} x_w, & |x_w| \geq \text{TH}, \\ 2x_w - \text{TH}, & \text{TH}/2 \leq x_w < \text{TH}, \\ \text{TH} + 2x_w, & -\text{TH} < x_w \leq -\text{TH}/2, \\ 0, & |x_w| < \text{TH}/2, \end{cases} \tag{56}$$

where TH denotes the threshold value and x_w represents the coefficients of the high frequency components of the DWT.

An enhanced version of the noisy speech signal using the wavelet hard thresholding method with one level decomposition is shown in Fig. 18. The effect of the wavelet hard thresholding method, with one level decomposition, on the speaker identification process is shown in Fig. 19. An enhanced version of the noisy speech signal using the wavelet soft thresholding method with one level decomposition is shown in Fig. 20. The effect of the wavelet soft thresholding

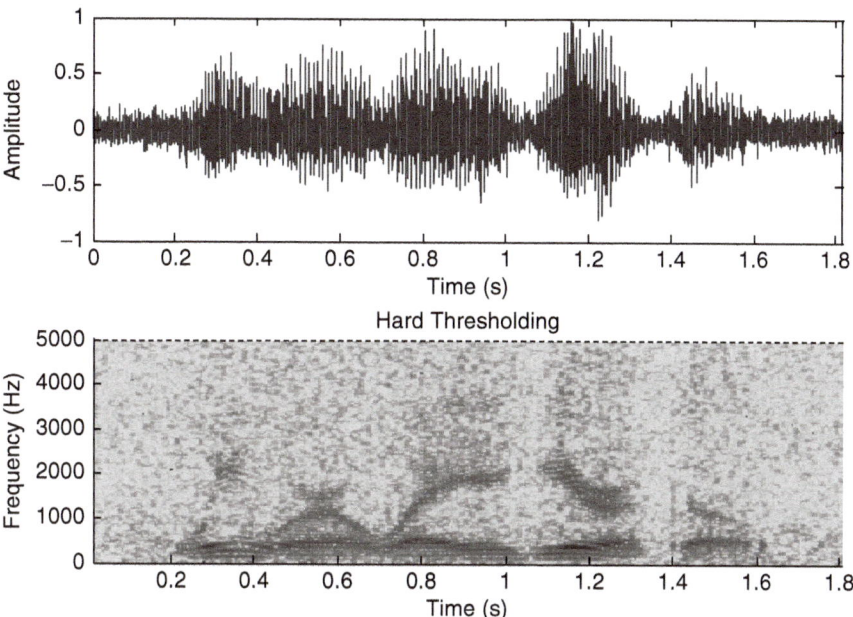

Fig. 18 Time domain waveform and spectrogram of the enhanced signal using the wavelet hard thresholding method with 1 level Haar wavelet transform. SNR = 6.5002 dB, SNRseg = 6.4605 dB, LLR = 0.1945, SD = 7.6423 dB

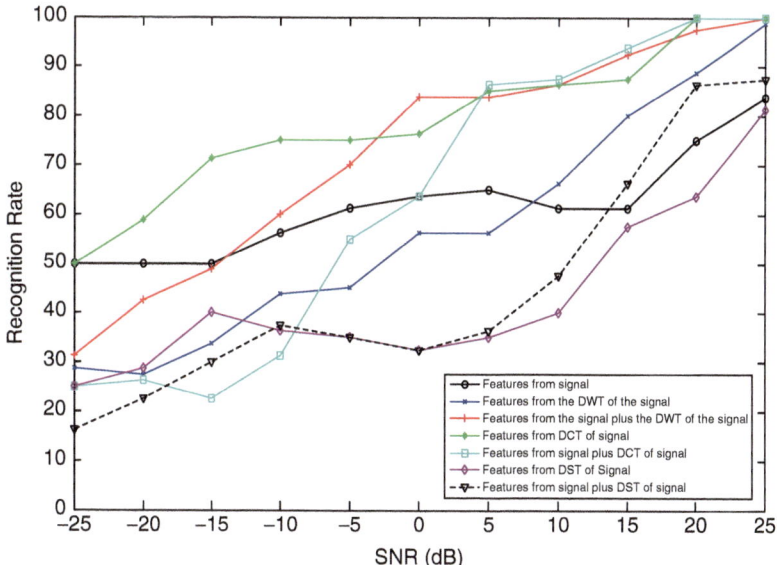

Fig. 19 Recognition rate vs. SNR for the different feature extraction methods in the presence of AWGN using the wavelet hard thresholding method with 1 level Haar wavelet transform

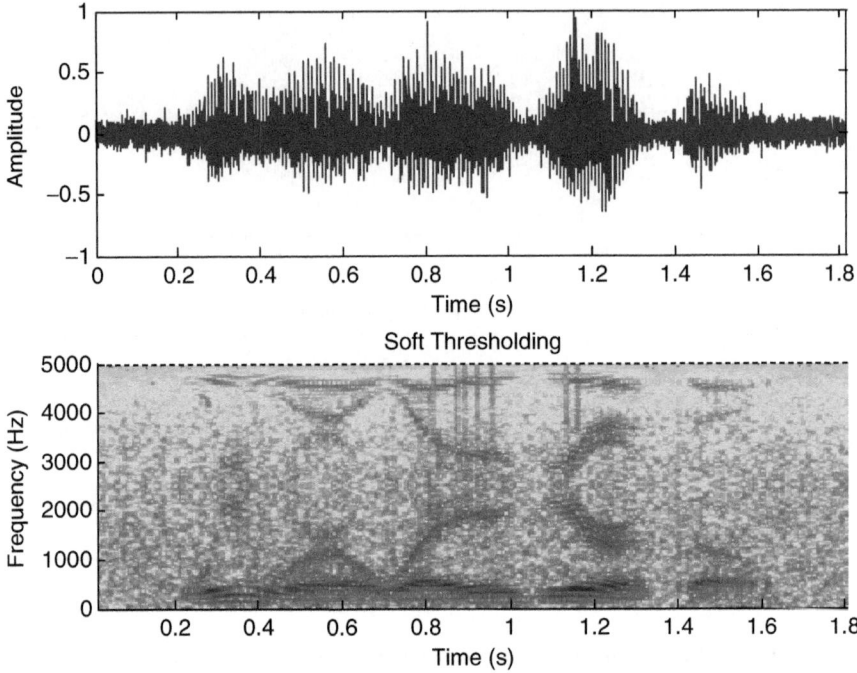

Fig. 20 Time domain waveform and spectrogram of the enhanced signal using the wavelet soft thresholding method with 1 level Haar wavelet transform. SNR = 6.4884 dB, SNRseg = 6.4506 dB, LLR = 0.1942, SD = 7.6463 dB

method, with one level decomposition, on the speaker identification process is shown in Fig. 21. The effect of the wavelet hard and soft thresholding methods, with two levels decomposition, on the speaker identification process is shown in Figs. 22 and 23, respectively. From these figures, it is clear that the wavelet denoising has the best effect on the speaker identification process. Soft thresholding with two levels wavelet thresholding gives the highest recognition rates. Thus, the wavelet denoising can be used with speaker identification systems implementing features extracted from the DCT of signals to get the highest recognition rates in noisy environments.

5 Blind Signal Separation

Blind signal separation can be used to reduce interference with undesired signals prior to the speaker identification process. In some cases, the speakers to be identified give utterances that are contaminated by noise or some kind of interference. Blind signal separation can be used for the separation of required speech

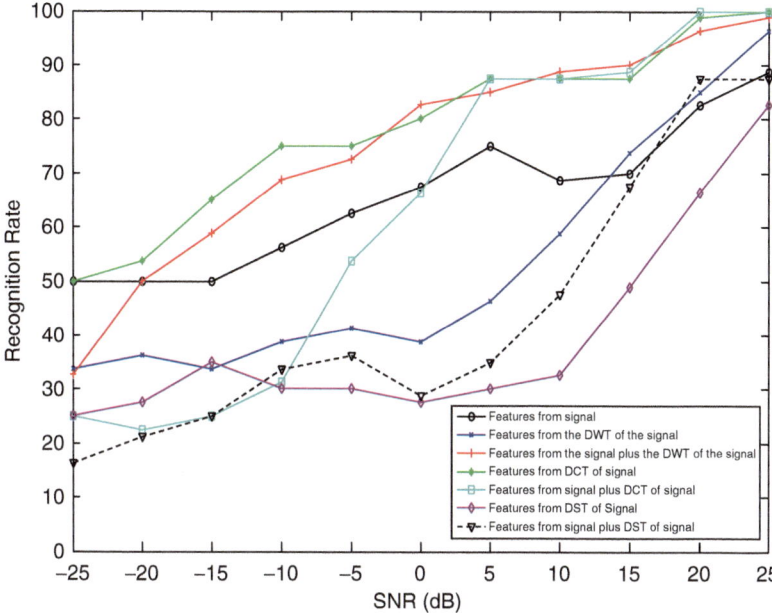

Fig. 21 Recognition rate vs. SNR for the different feature extraction methods in the presence of AWGN using the wavelet soft thresholding method with 1 level Haar wavelet transform

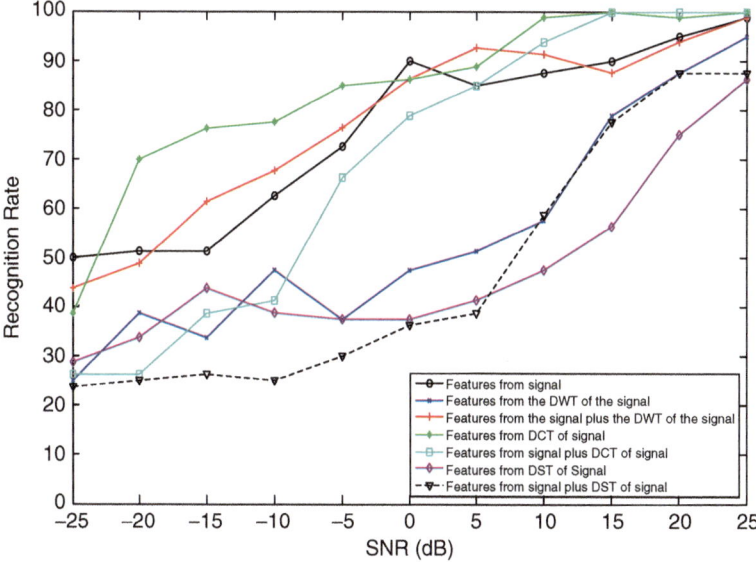

Fig. 22 Recognition rate vs. SNR for the different feature extraction methods in the presence of AWGN using the wavelet hard thresholding method with 2 levels Haar wavelet transform

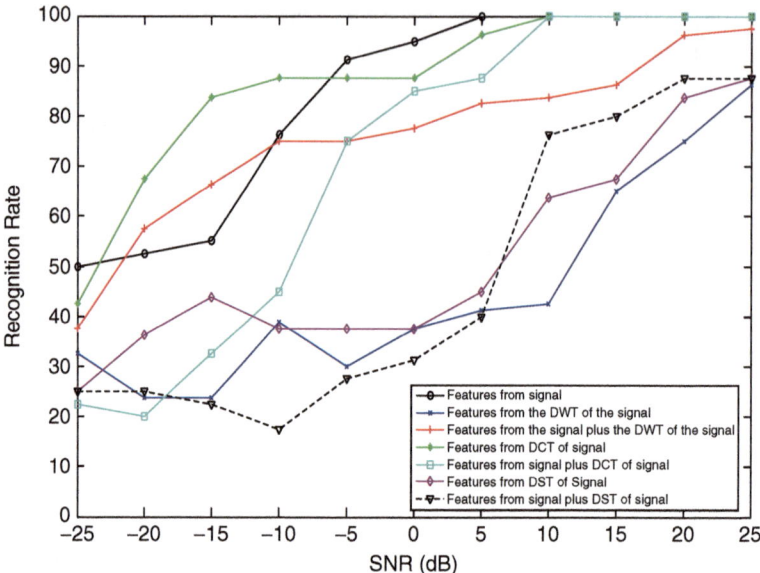

Fig. 23 Recognition rate vs. SNR for the different feature extraction methods in the presence of AWGN using the wavelet soft thresholding method with 2 levels Haar wavelet transform

signals from the background signals, and then the features can be extracted from the separated speech signals.

Blind signal separation deals with mixtures of signals in the presence of noise. If there are two original signals $s_1(n)$ and $s_2(n)$, which are mixed to give two observations $x_1(n)$ and $x_2(n)$, these observations can be represented as follows [50, 51]:

$$x_1(n) = \sum_{i=0}^{p} h_{11}(i)s_1(n-i) + \sum_{i=0}^{p} h_{12}(i)s_2(n-i) + v_1(n),$$

$$x_2(n) = \sum_{i=0}^{p} h_{21}(i)s_1(n-i) + \sum_{i=0}^{p} h_{22}(i)s_2(n-i) + v_2(n) \tag{57}$$

or in matrix form as follows:

$$\begin{pmatrix} x_1(n) \\ x_2(n) \end{pmatrix} = \begin{pmatrix} \mathbf{h}_{11}^{\mathrm{T}} & \mathbf{h}_{12}^{\mathrm{T}} \\ \mathbf{h}_{21}^{\mathrm{T}} & \mathbf{h}_{22}^{\mathrm{T}} \end{pmatrix} \begin{pmatrix} \mathbf{s}_1(n) \\ \mathbf{s}_2(n) \end{pmatrix} + \begin{pmatrix} v_1(n) \\ v_2(n) \end{pmatrix}, \tag{58}$$

where

$$\mathbf{h}_{ij}^{\mathrm{T}} = [h_{ij}(0), \dots, h_{ij}(p)],$$

$$\mathbf{s}_i^{\mathrm{T}}(n) = [s_i(n), \dots, s_i(n-p)], \tag{59}$$

Fig. 24 A fully coupled
2×2 mixing system

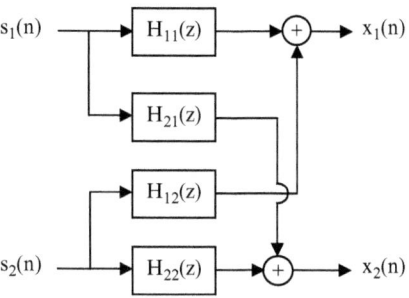

$v_1(n)$ and $v_2(n)$ are due to noise, $\mathbf{h}_{ij}$ is the impulse response from source j to sensor i, and p is the order of the filter. For simplicity, the source signals are assumed to be statistically independent with zero means. The problem is simplified by assuming that the signals arrive at the sensors unfiltered, which is equivalent to setting $\mathbf{h}_{11} = \mathbf{h}_{22} = 1$.

Taking Z-transform of (Eq. 58), and neglecting the effect of noise lead to:

$$\begin{pmatrix} X_1(z) \\ X_2(z) \end{pmatrix} = \begin{pmatrix} H_{11}(z) & H_{12}(z) \\ H_{21}(z) & H_{22}(z) \end{pmatrix} \begin{pmatrix} S_1(z) \\ S_2(z) \end{pmatrix}. \tag{60}$$

This model can be represented by the block diagram in Fig. 24. Simplifying (Eq. 60) leads to:

$$\begin{pmatrix} X_1(z) \\ X_2(z) \end{pmatrix} = \begin{pmatrix} 1 & H'_{21}(z) \\ H'_{21}(z) & 1 \end{pmatrix} \begin{pmatrix} S'_1(z) \\ S'_2(z) \end{pmatrix}, \tag{61}$$

where

$$\begin{aligned} S'_1(z) &= H_{11}(z) S_1(z), \\ S'_2(z) &= H_{22}(z) S_2(z), \\ H'_{12}(z) &= \frac{H_{12}(z)}{H_{22}(z)}, \\ H'_{21}(z) &= \frac{H_{21}(z)}{H_{11}(z)}. \end{aligned} \tag{62}$$

For $H_{ii}(z) = 1$, which is the case of interest, (Eq. 61) simplifies to:

$$\begin{pmatrix} X_1(z) \\ X_2(z) \end{pmatrix} = \begin{pmatrix} 1 & H_{12}(z) \\ H_{21}(z) & 1 \end{pmatrix} \begin{pmatrix} S_1(z) \\ S_2(z) \end{pmatrix}. \tag{63}$$

The objective of blind signal separation is to get the signals $y_1(n)$ and $y_2(n)$, which are as close as possible to $s_1(n)$ and $s_2(n)$. We can assume that:

$$\begin{pmatrix} Y_1(z) \\ Y_2(z) \end{pmatrix} = \begin{pmatrix} 1 & W_1(z) \\ W_2(z) & 1 \end{pmatrix} \begin{pmatrix} X_1(z) \\ X_2(z) \end{pmatrix}, \tag{64}$$

where in vector form

$$\begin{aligned} \mathbf{w}_i^T &= [w_i(0), \dots, w_i(q)], \\ \mathbf{x}_i^T(n) &= [x_i(n), \dots, x_i(n-q)]. \end{aligned} \tag{65}$$

Substituting (Eq. 63) into (Eq. 64) leads to [50, 51]:

$$\begin{pmatrix} Y_1(z) \\ Y_2(z) \end{pmatrix} = \begin{pmatrix} 1 + W_1(z)H_{21}(z) & W_1(z) + H_{12}(z) \\ W_2(z) + H_{21}(z) & 1 + W_2(z)H_{12}(z) \end{pmatrix} \begin{pmatrix} S_1(z) \\ S_2(z) \end{pmatrix}. \tag{66}$$

The time domain iterative separation algorithm for the 2×2 convolutive system minimizes the output cross-correlations for an arbitrary number of lags with $\delta + 1$ tap FIR filters. From (Eq. 66), it is clear that the solution of the problem is to find suitable $W_1(z)$ and $W_2(z)$, such that each of $Y_1(z)$ and $Y_2(z)$ contains only $S_1(z)$ or $S_2(z)$. This is achieved only if either the diagonal or the anti-diagonal elements of the cross-correlation matrices are zeros. Figure 25 shows a block diagram of the separation algorithm.

Assuming $s_1(n)$ and $s_2(n)$ are stationary, zero mean and independent random signals, the cross-correlation between the two signals is equal to zero, that is [50, 51]:

$$r_{s_1 s_2}(l) = E[s_1(n)s_2(n+l)] = 0 \quad \forall \quad l. \tag{67}$$

If each of $y_1(n)$ and $y_2(n)$ contains components of $s_1(n)$ or $s_2(n)$ only, then the cross-correlation between $y_1(n)$ and $y_2(n)$ should also be zero as follows:

$$r_{y_1 y_2}(l) = E[y_1(n)y_2(n+l)] = 0 \quad \forall \quad l. \tag{68}$$

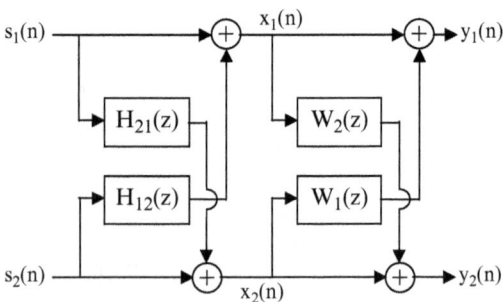

Fig. 25 Schematic diagram of the 2×2 separation algorithm

Substituting (Eq. 59) into (Eq. 68) gives:

$$r_{y_1 y_2}(l) = E[(x_1(n) + \mathbf{w}_1^T \mathbf{x}_2(n))(x_2(n+l) + \mathbf{w}_2^T \mathbf{x}_1(n+l))]. \tag{69}$$

If $r_{x_i x_j}(l) = E[x_i(n) x_j(n+l)]$, (Eq. 69) becomes:

$$r_{y_1 y_2}(l) = r_{x_1 x_2}(l) + \mathbf{w}_1^T \begin{pmatrix} r_{x_2 x_2}(l) \\ \vdots \\ r_{x_2 x_2}(l+q) \end{pmatrix} + \mathbf{w}_2^T \begin{pmatrix} r_{x_1 x_1}(l) \\ \vdots \\ r_{x_1 x_1}(l+q) \end{pmatrix} + \mathbf{w}_1^T \mathbf{R}_{x_2 x_1}(l) \mathbf{w}_2, \tag{70}$$

where $\mathbf{R}_{x_2 x_1}(l) = E[\mathbf{x}_2(n)(\mathbf{x}_1(n+l))^T]$ is a $(\delta + 1) \times (\delta + 1)$ matrix, which is a function of the cross-correlation between $\mathbf{x}_1$ and $\mathbf{x}_2$.

The cost function C is defined as the sum of the squares of the cross-correlations between the two inputs as follows [50, 51]:

$$C = \sum_{l=l_1}^{l_2} [r_{y_1 y_2}(l)]^2, \tag{71}$$

where l_1 and l_2 constitute a range of cross-correlation lags. C can also be expressed as:

$$C = \mathbf{r}_{y_1 y_2}^T \mathbf{r}_{y_1 y_2}, \tag{72}$$

where

$$\mathbf{r}_{y_1 y_2} = [r_{y_1 y_2}(l_1), \ldots, r_{y_1 y_2}(l_2)]^T. \tag{73}$$

Thus:

$$\mathbf{r}_{y_1 y_2} = \mathbf{r}_{x_1 x_2} + [\mathbf{Q}_{x_2 x_2}^+]^T \mathbf{w}_1 + [\mathbf{Q}_{x_1 x_1}^-]^T \mathbf{w}_2 + \mathbf{R}_{x_2 x_1}^T \mathbf{A}(\mathbf{w}_2) \mathbf{w}_1 \tag{74}$$

or

$$\mathbf{r}_{y_1 y_2} = \mathbf{r}_{x_1 x_2} + [\mathbf{Q}_{x_2 x_2}^+]^T \mathbf{w}_1 + [\mathbf{Q}_{x_1 x_1}^-]^T \mathbf{w}_2 + \mathbf{R}_{x_1 x_2}^T \mathbf{A}(\mathbf{w}_1) \mathbf{w}_2, \tag{75}$$

where $\mathbf{Q}_{x_2 x_2}^+$ and $\mathbf{Q}_{x_1 x_1}^-$ are $(\delta + 1) \times (l_2 - l_1 + 1)$ matrices, $\mathbf{R}_{x_2 x_1}$ is a $(2\delta + 1) \times (l_2 - l_1 + 1)$ matrix. These are all correlation matrices of x_1 and x_2 and are estimated using sample correlation estimates. $\mathbf{A}(\mathbf{w}_1)$ and $\mathbf{A}(\mathbf{w}_2)$ are $(2\delta + 1) \times (\delta + 1)$

matrices, which contain $\mathbf{w}_1$ and $\mathbf{w}_2$, respectively. In order to find some suitable $\mathbf{w}_1$ and $\mathbf{w}_2$, C is minimized such that:

$$\frac{\partial C}{\partial \mathbf{w}_\tau} = [0, \ldots, 0]^{\mathrm{T}}, \quad \tau = 1, 2. \tag{76}$$

Let

$$\begin{aligned} \psi_1 &= ([\mathbf{Q}_{x_2 x_2}^+]^{\mathrm{T}} + \mathbf{R}_{x_2 x_1}^{\mathrm{T}} \mathbf{A}(\mathbf{w}_2)), \\ \psi_2 &= ([\mathbf{Q}_{x_1 x_1}^-]^{\mathrm{T}} + \mathbf{R}_{x_1 x_2}^{\mathrm{T}} \mathbf{A}(\mathbf{w}_1)). \end{aligned} \tag{77}$$

Substituting (Eq. 77) into (Eq. 74) and (Eq. 75) gives:

$$\mathbf{r}_{y_1 y_2} = \mathbf{r}_{x_1 x_2} + \psi_1 \mathbf{w}_1 + [\mathbf{Q}_{x_1 x_1}^-]^{\mathrm{T}} \mathbf{w}_2 \tag{78}$$

or

$$\mathbf{r}_{y_1 y_2} - \mathbf{r}_{x_1 x_2} + \psi_2 \mathbf{w}_2 + [\mathbf{Q}_{x_2 x_2}^+]^{\mathrm{T}} \mathbf{w}_1. \tag{79}$$

From (Eq. 76), we obtain [50, 51]:

$$\begin{aligned} \mathbf{w}_1 &= -(\psi_1^{\mathrm{T}} \psi_1)^{-1} \psi_1^{\mathrm{T}} (\mathbf{r}_{x_1 x_2} + [\mathbf{Q}_{x_1 x_1}^-]^{\mathrm{T}} \mathbf{w}_2), \\ \mathbf{w}_2 &= -(\psi_2^{\mathrm{T}} \psi_2)^{-1} \psi_2^{\mathrm{T}} (\mathbf{r}_{x_1 x_2} + [\mathbf{Q}_{x_2 x_2}^+]^{\mathrm{T}} \mathbf{w}_1). \end{aligned} \tag{80}$$

$\mathbf{w}_1$ and $\mathbf{w}_2$ are obtained by iterating between the two equations until convergence is achieved, when the rate of change of parameter values is less than a preset threshold. By estimating $\mathbf{w}_1$ and $\mathbf{w}_2$, we then obtain a set of outputs $y_1(n)$ and $y_2(n)$. Each output contains $s_1(n)$ or $s_2(n)$, only.

The above-mentioned blind signal separation algorithm can be applied on the signal mixtures in time domain or in a transform domain such as the DCT or the DST. Wavelet denoising can also be used for noise reduction in the resulting separated signals. In the DCT or DST transform domains, the separation is performed on a few coefficients in the transform domain due to the energy compaction property. Figures 26–37 confirm the superiority of transform domain separation to time domain separation and the importance of the wavelet denoising step for two mixtures composed of speech and music signals in the presence of noise.

The effect of blind signal separation on the performance of speaker identification systems is shown in Fig. 38. This figure reveals that signal separation for the desired speech signals is very important for robust speaker identification because interfering signals at low SNRs destroy the distinguishing features of speech signals.

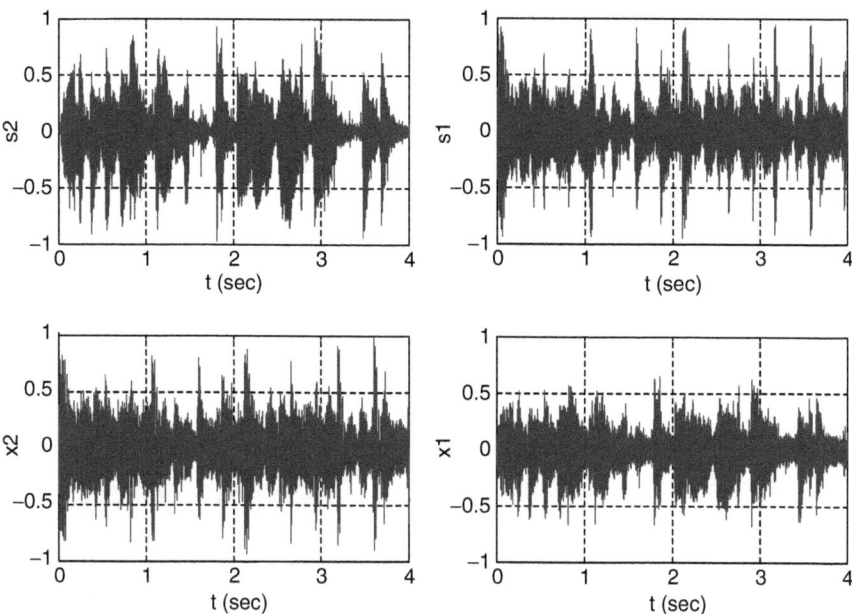

Fig. 26 Original signals and noisy mixtures. (**a**) Original speech signal. (**b**) Original music signal. (**c**) Noisy mixture 1, SNR = −10 dB. (**d**) Noisy mixture 2, SNR = −10 dB

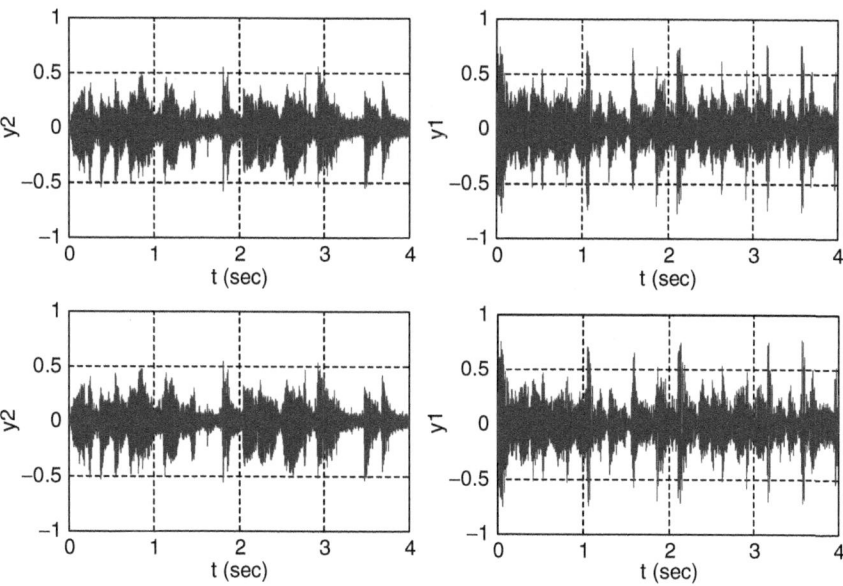

Fig. 27 Signal separation in the time domain with and without wavelet denoising. (**a**) Separated speech signal in the absence of wavelet denoising. SNR = −8.31 dB, SNRseg =−8.33 dB, LLR = 0.45, SD = 22.88 dB. (**b**) Separated music signal in the absence of wavelet denoising. SNR = −3.24 dB, SNRseg = −3.28 dB, LLR = 0.52, SD = 15.64 dB. (**c**) Separated speech signal in the presence of wavelet denoising. SNR = −5.43 dB, SNRseg = −5.64 dB, LLR = 0.47, SD = 18.20 dB. (**d**) Separated music signal in the presence of wavelet denoising. SNR = −0.49 dB, SNRseg = −0.67 dB, LLR = 0.54, SD = 12.80 dB

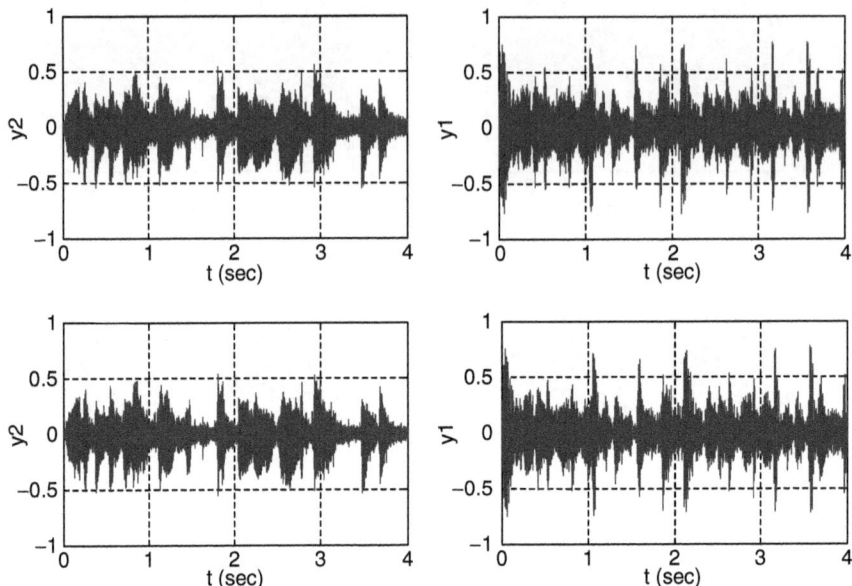

Fig. 28 Signal separation using the DCT with and without wavelet denoising. (**a**) Separated speech signal in the absence of wavelet denoising. SNR = −7.72 dB, SNRseg = −7.74 dB, LLR = 0.44, SD = 21.85 dB. (**b**) Separated music signal in the absence of wavelet denoising. SNR = 2.08 dB, SNRseg = 2.05 dB, LLR = 0.51, SD = 11.01 dB. (**c**) Separated speech signal in the presence of wavelet denoising. SNR = −4.87 dB, SNRseg = −5.08 dB, LLR = 0.48, SD = 17.42 dB. (**d**) Separated music signal in the presence of wavelet denoising. SNR = 4.01 dB, SNRseg = 3.94 dB, LLR = 0.42, SD = 10.06 dB

6 Deconvolution of Speech Signals

Deconvolution methods can be used in a preprocessing step in the testing phase of the speaker identification system to eliminate the channel degradation effect as shown in Fig. 39.

6.1 LMMSE Deconvolution

The linear shift invariant speech degradation model for a speech signal that passed through a finite bandwidth channel can be described as a convolution between the signal and the channel impulse response in the presence of noise. This convolution can be put in matrix vector notation as follows [52–55]:

$$\mathbf{x} = \mathbf{Hs} + \mathbf{v}, \tag{81}$$

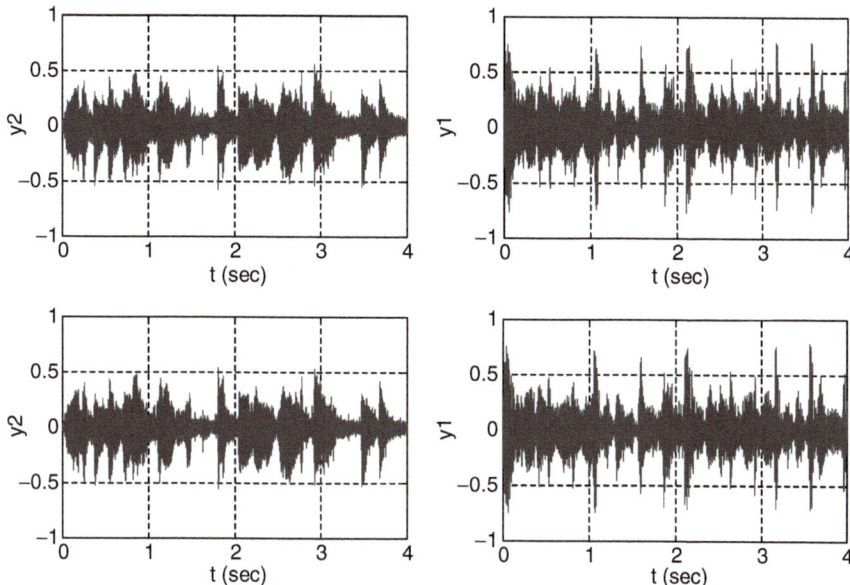

Fig. 29 Signal separation using the DST with and without wavelet denoising. (**a**) Separated speech signal in the absence of wavelet denoising. SNR = −7.72 dB, SNRseg = −7.74 dB, LLR = 0.44, SD = 21.85 dB. (**b**) Separated music signal in the absence of wavelet denoising. SNR = 2.08 dB, SNRseg = 2.05 dB, LLR = 0.51, SD = 11.01 dB. (**c**) Separated speech signal in the presence of wavelet denoising. SNR = −4.87 dB, SNRseg = −5.08 dB, LLR = 0.48, SD = 17.42 dB. (**d**) Separated music signal in the presence of wavelet denoising. SNR = 4.01 dB, SNRseg = 3.95 dB, LLR = 0.42, SD = 10.06 dB

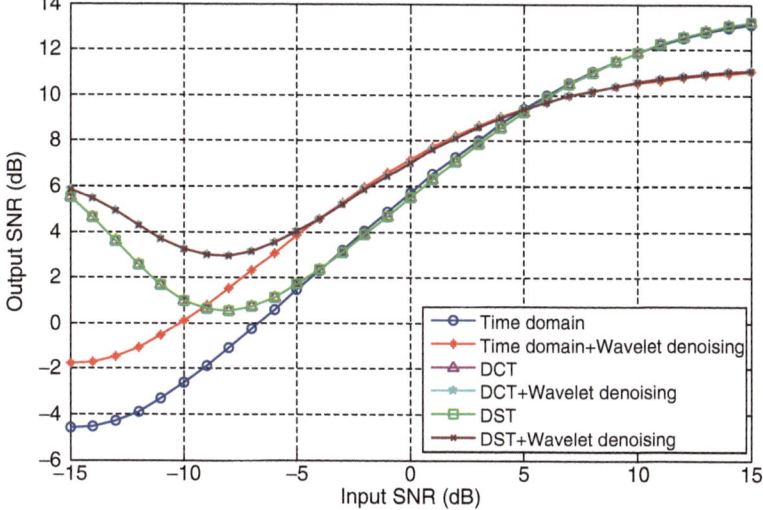

Fig. 30 Output SNR of the music signal vs. input SNR for all separation methods

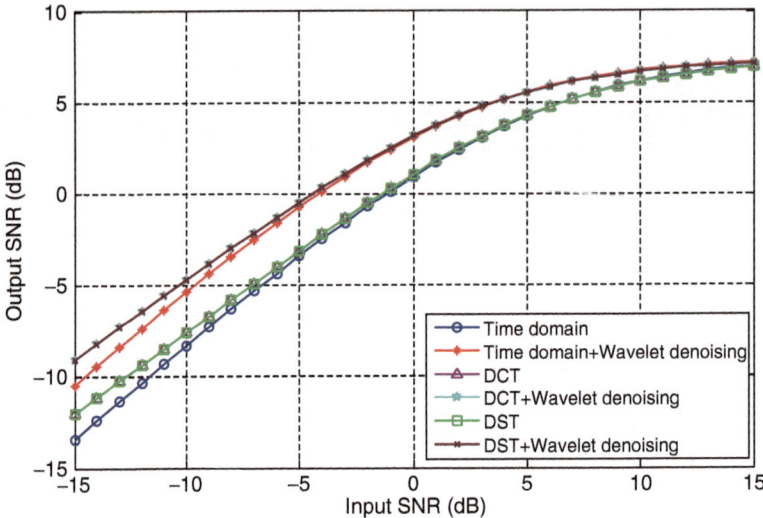

Fig. 31 Output SNR of the speech signal vs. input SNR for all separation methods

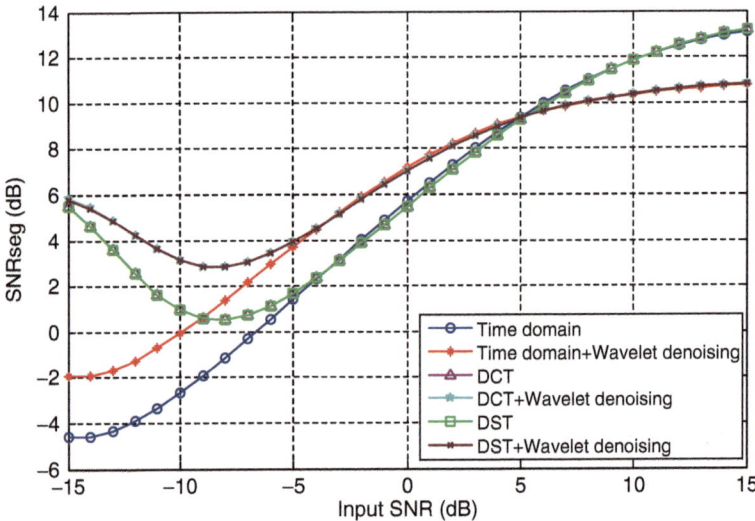

Fig. 32 Output SNRseg of the music signal vs. input SNR for all separation methods

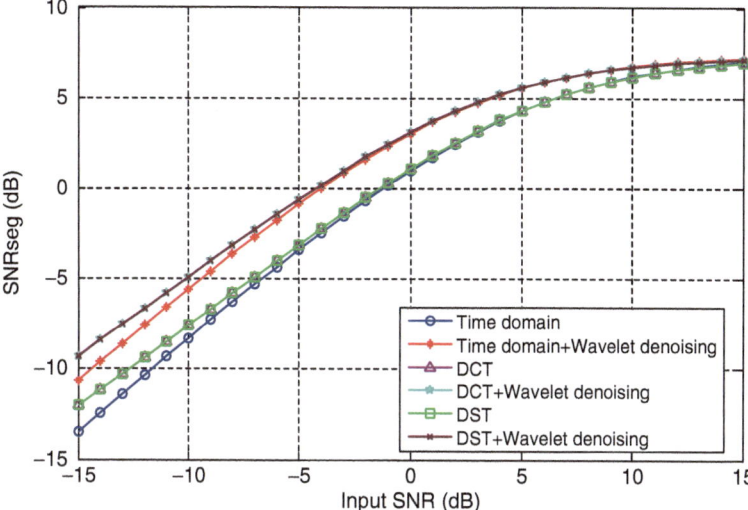

Fig. 33 Output SNRseg of the speech signal vs. input SNR for all separation methods

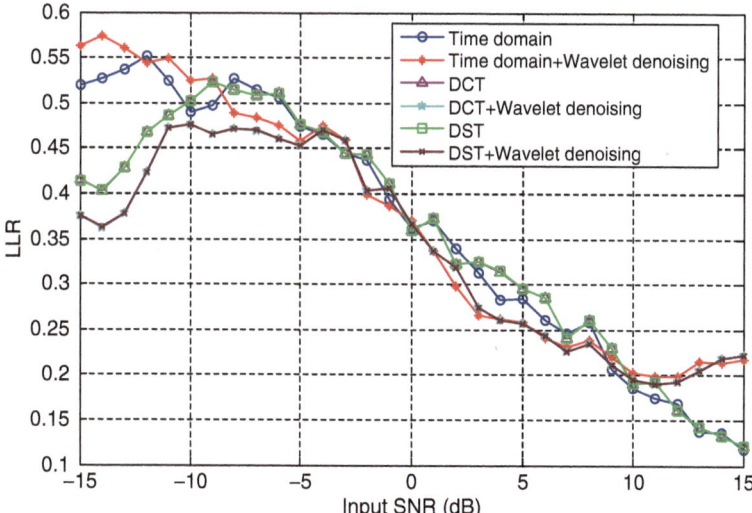

Fig. 34 Output LLR of the music signal vs. input SNR for all separation methods

where $\mathbf{s}$, $\mathbf{x}$, and $\mathbf{v}$ are vectors of length N, of the original speech signal, the degraded speech signal, and the noise, respectively. The matrix $\mathbf{H}$ is the $N \times N$ channel matrix. For a linear shift invariant system, the matrix $\mathbf{H}$ is a block Toeplitz matrix.

The problem is to estimate $\mathbf{s}$ given the recorded speech $\mathbf{x}$. It is required that the MSE of estimation be minimum over the entire ensemble of all possible estimates of the speech signal [55, 56].

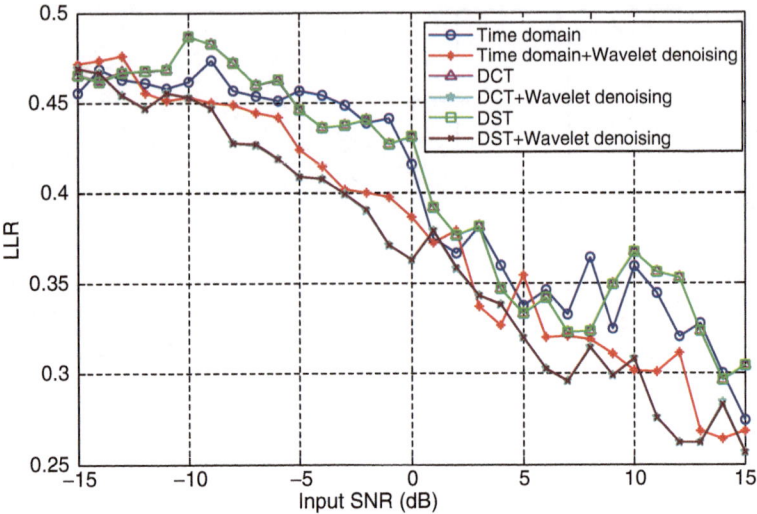

Fig. 35 Output LLR of the speech signal vs. input SNR for all separation methods

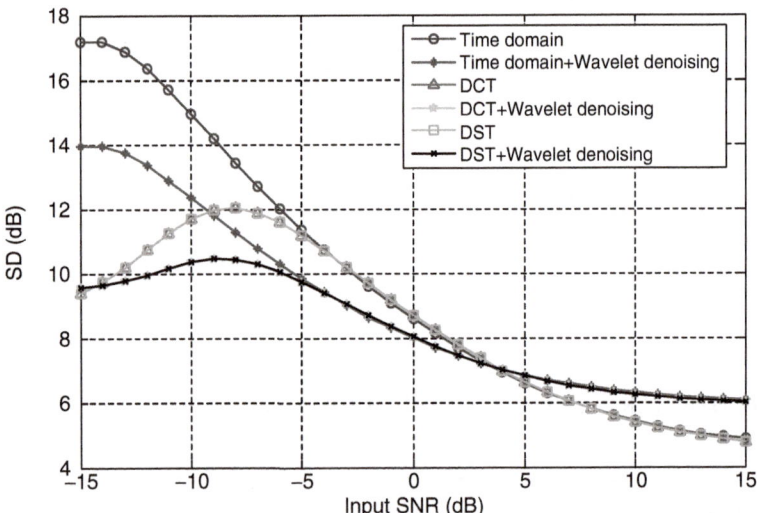

Fig. 36 Output SD of the music signal vs. input SNR for all separation methods

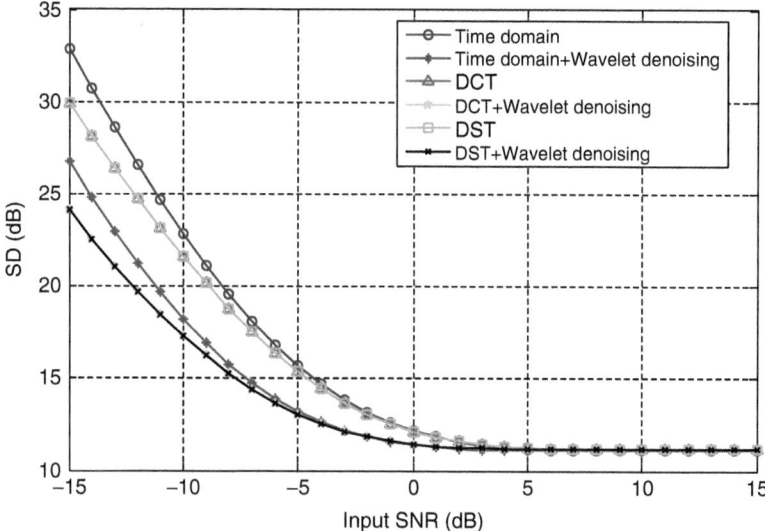

Fig. 37 Output SD of the speech signal vs. input SNR for all separation methods

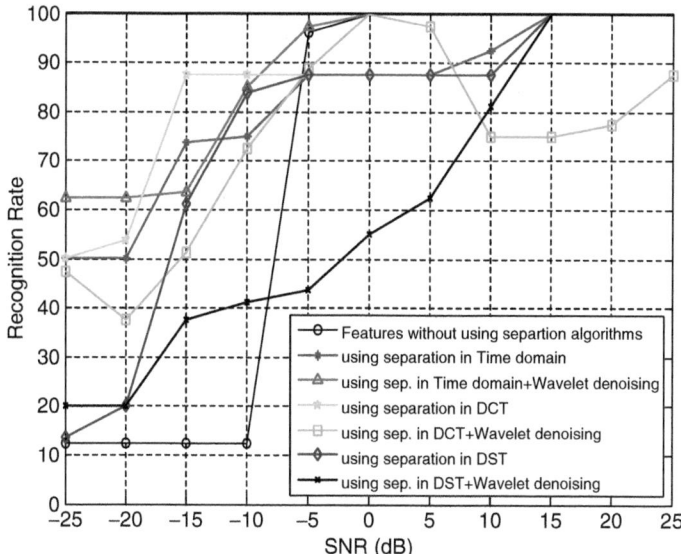

Fig. 38 Recognition rate vs. SNR for the different feature extraction methods in the presence of AWGN with blind signal separation as a preprocessing step

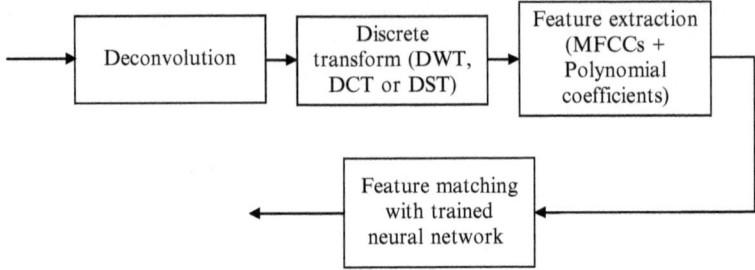

Fig. 39 Speaker identification with deconvolution

$$\min_{\hat{x}} E[\boldsymbol{\varepsilon}^t\boldsymbol{\varepsilon}] = \mathbf{E}[\mathrm{Tr}(\boldsymbol{\varepsilon}\boldsymbol{\varepsilon}^t)], \qquad (82)$$

where $\boldsymbol{\varepsilon} = \mathbf{s} - \hat{\mathbf{s}}$ is the estimation error and $\hat{\mathbf{s}}$ is an estimate of original speech signal.

Since the transformation matrix $\mathbf{H}$ is linear, the estimate of $\mathbf{s}$ will be linear. That is an estimate of $\mathbf{s}$ that can be derived by a linear operation on the degraded speech signal as follows:

$$\hat{\mathbf{s}} = \mathbf{Lx}, \qquad (83)$$

where $\mathbf{L}$ is the derived subject to solving (Eq. 82), which leads to the following equation:

$$\begin{aligned}
\min_{\hat{x}} E[\mathrm{Tr}(\boldsymbol{\varepsilon}\boldsymbol{\varepsilon}^t)] &= E[\mathrm{Tr}\{(\mathbf{s} - \mathbf{Lx})(\mathbf{s} - \mathbf{Lx})^t\}] \\
&= E[\mathrm{Tr}\{\mathbf{ss}^t - \mathbf{L}(\mathbf{Hss}^t + \mathbf{vs}^t) - (\mathbf{ss}^t\mathbf{H}^t + \mathbf{sv}^t)\mathbf{L}^t \\
&\quad + \mathbf{L}(\mathbf{Hss}^t\mathbf{H}^t + \mathbf{vs}^t\mathbf{H}^t + \mathbf{Hsv}^t + \mathbf{vv}^t)\mathbf{L}^t\}].
\end{aligned} \qquad (84)$$

We have $\mathrm{Tr}(\mathbf{A}) = \mathrm{Tr}(\mathbf{A}^t)$. Since the trace is linear, it can be interchanged with the expectation operator. Equation 84 can be simplified using some assumptions. The noise is assumed to be independent of the speech signal. This assumption leads to:

$$E[\mathbf{sv}^t] = E[\mathbf{v}^t\mathbf{s}] = [\mathbf{0}]. \qquad (85)$$

The autocorrelation matrices can be defined as:

$$E[\mathbf{ss}^t] = \mathbf{R}_s, \qquad (86)$$

$$E[\mathbf{vv}^t] = \mathbf{R}_v. \qquad (87)$$

Substituting from (Eq. 85), (Eq. 86) and (Eq. 87) into (Eq. 84) yields:

$$\min_{\hat{s}} E[\mathrm{Tr}(\boldsymbol{\varepsilon\varepsilon}^t)] = \mathrm{Tr}\{\mathbf{R_s} - 2\mathbf{LHR_s} + \mathbf{LHR_sH^tL^t} + \mathbf{LR_vL^t}\}. \qquad (88)$$

Differentiating (Eq. 88) with respect to $\mathbf{L}$ and setting the result equal to zero, the LMMSE solution is given by:

$$\mathbf{L} = \mathbf{R_sH^t(HR_sH^t + R_v)^{-1}}. \qquad (89)$$

The solution to obtain the estimate $\hat{s}$ requires the tedious task of inverting an $N \times N$ matrix. This task can be avoided using the Toeplitz-to-circulant approximation of matrices [57, 58].

Mathematical operations on matrices are greatly simplified, when these matrices have circulant structures. The simplifications emerge from the fact that operations on circulant matrices yield circulant matrices. Circulant matrices can be classified to either circulant or block circulant matrices. Both types of matrices can be diagonalized via either the 1D or the 2D DFT. This attractive property allows the inversion of circulant matrices of large dimensions, since the inversion process will be applied to diagonal sparse matrices [57, 58].

Let $\mathbf{Q}$ be an $S \times S$ Toeplitz matrix of the following form [57, 58]:

$$\mathbf{Q} = \begin{bmatrix} q(0) & \cdots & q(-\Pi) & & 0 \\ \vdots & \ddots & & \ddots & \\ q(\Gamma) & & \ddots & & q(-\Pi) \\ & \ddots & & \ddots & \vdots \\ 0 & & q(\Gamma) & \cdots & q(0) \end{bmatrix}. \qquad (90)$$

It can be approximated by an $S \times S$ circulant matrix $\mathbf{Q^c}$ defined as [57, 58]:

$$\mathbf{Q^c} = \begin{bmatrix} q(0) & \cdots & \cdots & q(-\Pi) & 0 & \cdots & q(\Gamma) & \cdots & q(1) \\ \vdots & \ddots & & & \ddots & \ddots & \cdots & \ddots & \\ \vdots & & \ddots & & & \ddots & & \ddots & \cdots & q(\Gamma) \\ q(\Gamma) & & & \ddots & & \ddots & & \ddots & \cdots \\ 0 & \ddots & & & \ddots & & & \ddots & 0 \\ \vdots & \ddots & \ddots & & & \ddots & & & q(-\Pi) \\ q(-\Pi) & \vdots & \ddots & \ddots & & & \ddots & & \vdots \\ \vdots & \ddots & \vdots & \ddots & \ddots & & & \ddots & \vdots \\ q(-1) & \cdots & q(-\Pi) & \vdots & 0 & q(\Gamma) & \cdots & \cdots & q(0) \end{bmatrix}, \qquad (91)$$

where each row is a circular shift of the row above, and the first row is a circular shift of the last row. The primary difference between the matrices $\mathbf{Q}$ and $\mathbf{Q}^c$ is the elements added at the upper right and lower left corners to produce the cyclic structure in the rows. If the matrix size is large and the number of nonzero elements on the main diagonals compared to the number of zero elements is small (i.e., the matrix is sparse), the elements added at the upper right and lower left corners do not affect the matrix, because the number of these elements is small compared to the number of the main diagonal elements. It can be shown from the Eigenvalues distribution of the matrices $\mathbf{Q}$ and $\mathbf{Q}^c$ that both matrices are asymptotically equivalent.

It is known that an $S \times S$ circulant matrix $\mathbf{Q}^c$ is diagonalized as follows [57, 58]:

$$\Lambda = \boldsymbol{\varphi}^{-1} \mathbf{Q}^c \boldsymbol{\varphi}, \tag{92}$$

where Λ is an $S \times S$ diagonal matrix, whose elements $\lambda(s, s)$ are the Eigenvalues of $\mathbf{Q}^c$, and $\boldsymbol{\varphi}$ is an $S \times S$ unitary matrix of the Eigenvectors of $\mathbf{Q}^c$. Thus, we have:

$$\boldsymbol{\varphi}\boldsymbol{\varphi}^{*t} = \boldsymbol{\varphi}^{*t}\boldsymbol{\varphi} = \mathbf{I}. \tag{93}$$

The elements $\varphi(s_1, s_2)$ of $\boldsymbol{\varphi}$ are given by:

$$\varphi(s_1, s_2) = e^{j2\pi s_1 s_2/S} \tag{94}$$

for $s_1, s_2 = 0, 1, \ldots, S - 1$.

The Eigenvalues $\lambda(s, s)$ can be referred to as $\lambda(s)$. For these Eigenvalues, the following relation holds:

$$\lambda(s) = q(0) + \sum_{m=1}^{\Gamma} q(m)e^{-j2\pi ms/S} + \sum_{m=-\Pi}^{-1} q(m)e^{-j2\pi ms/S}, \tag{95}$$

$$s = 0, 1, \ldots, S - 1.$$

Because of the cyclic nature of $\mathbf{Q}^c$, we can define:

$$q(S - m) = q(-m). \tag{96}$$

Thus, (Eq. 95) can be written in the form:

$$\lambda(s) = \sum_{m=0}^{S-1} q(m)e^{-j2\pi ms/S}, \tag{97}$$

where $s = 0, 1, \ldots, S - 1$.

Thus, the circulant matrix can be simply diagonalized by computing the DFT of the cyclic sequence $q(0), q(1), \ldots, q(S - 1)$.

The implementation of the LMMSE deconvolution method is greatly dependent on the Toeplitz-to-circulant approximation. As mentioned above, the solution of

(Eq. 89) requires the inversion of an $N \times N$ matrix. To avoid this process, we can benefit from the Toeplitz-to-circulant approximation.

First consider the 1D Fourier transform of the speech signal $s(n)$ given in (Eq. 6). This equation can also be written in vector-matrix notation as:

$$\mathbf{S} = \boldsymbol{\varphi}^{-1}\mathbf{s}, \tag{98}$$

where $\mathbf{S}$ and $\mathbf{s}$ are matrix vectors of $S(k)$ and $s(n)$, respectively. The $N \times N$ matrix $\boldsymbol{\varphi}^{-1}$ contains the complex exponentials of the 1D Fourier transform. Similarly, the 1D IDFT can be obtained by multiplying both sides of (Eq. 98) by the matrix $\boldsymbol{\varphi}$, which yields:

$$\mathbf{s} = \boldsymbol{\varphi}\mathbf{S}. \tag{99}$$

From the Toeplitz structures of $\mathbf{H}$, $\mathbf{R_s}$, and $\mathbf{R_v}$, which can be approximated by circulant matrices, we get [57, 58]:

$$\boldsymbol{\varphi}^{-1}\hat{\mathbf{s}} = \boldsymbol{\varphi}^{-1}\mathbf{R_s}\boldsymbol{\varphi}\boldsymbol{\varphi}^{-1}\mathbf{H}^t\boldsymbol{\varphi}\boldsymbol{\varphi}^{-1}[\mathbf{HR_xH}^t + \mathbf{R_n}]^{-1}\boldsymbol{\varphi}\boldsymbol{\varphi}^{-1}\mathbf{x}. \tag{100}$$

The above equation leads to:

$$\boldsymbol{\varphi}^{-1}\hat{\mathbf{s}} = \boldsymbol{\varphi}^{-1}\mathbf{R_s}\boldsymbol{\varphi}\boldsymbol{\varphi}^{-1}\mathbf{H}^t\boldsymbol{\varphi}[\boldsymbol{\varphi}^{-1}\mathbf{H}\boldsymbol{\varphi}\boldsymbol{\varphi}^{-1}\mathbf{R_x}\boldsymbol{\varphi}\boldsymbol{\varphi}^{-1}\mathbf{H}^t\boldsymbol{\varphi} + \boldsymbol{\varphi}^{-1}\mathbf{R_n}\boldsymbol{\varphi}]^{-1}\boldsymbol{\varphi}^{-1}\mathbf{x}. \tag{101}$$

Using the diagonalization property, the following form is obtained:

$$\hat{\mathbf{S}} = \Lambda_s\Lambda_h^*[\Lambda_h\Lambda_s\Lambda_h^* + \Lambda_v]^{-1}\mathbf{X}, \tag{102}$$

where $\Lambda_s = \boldsymbol{\varphi}^{-1}\mathbf{R_s}\boldsymbol{\varphi}$, and $\Lambda_h = \boldsymbol{\varphi}^{-1}\mathbf{H}\boldsymbol{\varphi}$ are diagonal matrices whose elements are the Eigenvalues of the matrices $\mathbf{R_s}$ and $\mathbf{H}$, respectively. Λ_h^* is a diagonal matrix, whose elements are the complex conjugates of the elements of Λ_h. $\hat{\mathbf{S}} = \boldsymbol{\varphi}^{-1}\hat{\mathbf{s}}$ and $\mathbf{X} = \boldsymbol{\varphi}^{-1}\mathbf{x}$ represent the 1D DFT of the estimated and degraded speech signals, respectively. This diagonalization process allows the operation on sparse matrices, which can be inverted easily.

The Eigenvalues of the matrix $\mathbf{R_s}$ are obtained from the 1D DFT of the correlation sequence $R_s(n)$, which represents the circular sequence of the matrix $\mathbf{R_s}$. Also the Eigenvalues of the matrix $\mathbf{H}$ are obtained from the 1D DFT of the channel impulse response sequence.

Another problem encountered in the LMMSE deconvolution method is how to estimate the correlation sequence $R_s(n)$ of the original speech signal. This correlation sequence can be estimated from a prototype speech signal $s'(n)$ using the following equation [55]:

$$R_s(n) \cong \frac{1}{w}\sum_{l=1}^{w} s'(l)s'(n+l), \tag{103}$$

where $R_s(n)$ is the correlation at index n and w is an arbitrary window length. The prototype speech signal $s'(n)$ may be taken as the degraded speech signal $x(n)$. Thus, the correlation sequence may be approximated from the degraded speech signal as [55]:

$$R_s(n) \cong \frac{1}{w} \sum_{l=1}^{w} x(l)x(n+l). \tag{104}$$

6.2 Inverse Filter Deconvolution

The speech deconvolution problem can be solved directly by inverting the channel impulse response operator. This direct deconvolution method is feasible in the absence of noise, but severe distortions are observed in the restored speech signals at low SNRs.

The mathematical model for the inverse filter deconvolution method is based on assuming a known and invertible channel impulse response operator. A direct solution to the deconvolution problem can be obtained by estimating $\hat{s}$ that minimizes the norm of the difference between the reconvolved estimated speech signal $H\hat{s}$ and the degraded speech x. Mathematically, this can be represented by estimating $\hat{s}$ that minimizes the following cost function [57, 58]:

$$\Psi(\hat{s}) = ||x - H\hat{s}||^2. \tag{105}$$

Taking the partial derivative of the both sides of (Eq. 105) with respect to $\hat{s}$ and setting it equal to zero yields:

$$\frac{\partial \Psi(\hat{s})}{\partial \hat{s}} = 0 = -2H^t[x - H\hat{s}]. \tag{106}$$

This leads to:

$$\hat{s} = [H^tH]^{-1}H^tx. \tag{107}$$

The above equation can be simplified to the form:

$$\hat{s} = H^{-1}x = s + H^{-1}v. \tag{108}$$

The Toeplitz-to-circulant approximation is used to solve matrix inversion problem in (Eq. 108). Applying the operator φ^{-1} on both sides of the equation, yields:

$$\varphi^{-1}\hat{s} = \varphi^{-1}H^{-1}x = \varphi^{-1}H^{-1}\varphi\varphi^{-1}x = \varphi^{-1}H\varphi\varphi)^{-1}X. \tag{109}$$

Thus:

$$\hat{\mathbf{S}} = \Lambda_{\mathbf{h}}^{-1}\mathbf{X}. \tag{110}$$

In the above equation, the inversion process is performed on the diagonal matrix $\Lambda_{\mathbf{h}}$. So, it can be implemented easily due to the maximally sparse structure of this matrix.

Equation 110 can be written in an equivalent form as follows [57, 58]:

$$\hat{S}(k) = \frac{X(k)}{H(k)} = S(k) + \frac{V(k)}{H(k)}. \tag{111}$$

Thus, the deconvolution error can be expressed as:

$$\|\hat{S}(k) - S(k)\| = \left\|\frac{V(k)}{H(k)}\right\| = \sqrt{\sum_{k}\left|\frac{V(k)}{H(k)}\right|^2}. \tag{112}$$

The use of the inverse filter is limited to restoring noise free speech signals. This is due to the lowpass nature of the channel impulse response operator $\mathbf{H}$, which leads when inverted to the amplification of the high frequency noise components in the restored signal. This limitation is clear, especially when $\mathbf{H}$ is near singular. Thus, its inverse will have very large valued elements and consequently, the term $\mathbf{H}^{-1}\mathbf{v}$ can dominate the term containing the solution $\mathbf{s}$ in (Eq. 108). To overcome the limitation of the inverse filter deconvolution, some regularization is needed to avoid the amplification of the high frequency noise.

6.3 Regularized Deconvolution

An inverse problem is characterized as ill-posed, when there is no guarantee for the existence, uniqueness, and stability of the solution based on direct inversion. The solution of an inverse problem is not guaranteed to be stable if a small perturbation in the data can produce a large effect on the solution. Speech signal deconvolution belongs to a general class of ill-posed problems. Regularization theory, which was basically introduced by Tikhonov and Miller, provides a formal basis for the development of regularized solutions for ill-posed problems [59–61].

The stabilizing functional approach is one of the basic methodologies for the development of regularized solutions. According to this method, an ill-posed problem can be formulated as the constrained minimization of a certain functional, called stabilizing functional [81]. The specific constraints imposed by the stabilizing functional approach on the solution depend on the form and properties of the stabilizing functional used. From the nature of the problem, these constraints are necessarily related to the a priori information regarding the expected regularized solution.

According to the regularization method, the solution of (Eq. 81) is obtained by the minimization of the cost function [59–61]:

$$\Psi(\hat{\mathbf{s}}) = ||\mathbf{x} - \mathbf{H}\hat{\mathbf{s}}||^2 + \eta||\mathbf{C}\hat{\mathbf{s}}||^2, \tag{113}$$

where $\mathbf{C}$ is the regularization operator and η is the regularization parameter.

This minimization is accomplished by taking the derivative of the cost function yielding:

$$\frac{\partial \Psi(\hat{\mathbf{s}})}{\partial \hat{\mathbf{s}}} = \mathbf{0} = 2\mathbf{H}^t(\mathbf{x} - \mathbf{H}\hat{\mathbf{s}}) - 2\eta\mathbf{C}^t\mathbf{C}\hat{\mathbf{s}}. \tag{114}$$

Solving for $\hat{\mathbf{s}}$ that provides the minimum of the cost function yields:

$$\hat{\mathbf{s}} = [\mathbf{H}^t\mathbf{H} + \eta\mathbf{C}^t\mathbf{C}]^{-1}\mathbf{H}^t\mathbf{x} = \mathbf{A}(\eta)\mathbf{x}, \tag{115}$$

where

$$\mathbf{A}(\eta) = [\mathbf{H}^t\mathbf{H} + \eta\mathbf{C}^t\mathbf{C}]^{-1}\mathbf{H}^t. \tag{116}$$

The rule of the regularization operator $\mathbf{C}$ is to move the small Eigenvalues of $\mathbf{H}$ away from zero, while leaving the large Eigenvalues unchanged. The generality of the linear operator $\mathbf{C}$ allows the development of a variety of constraints that can be incorporated into the deconvolution operation. The simplest case, that will be considered in this book, is $\mathbf{C} = \mathbf{I}$. In this case the regularized solution reduces to the regularized inverse filter solution, which is named the pseudo inverse filter solution, and it is represented as:

$$\hat{\mathbf{s}} = [\mathbf{H}^t\mathbf{H} + \eta\mathbf{I}]^{-1}\mathbf{H}^t\mathbf{x}. \tag{117}$$

To perform the inversion process in (Eq. 117), the Toeplitz-to-circulant approximation is implemented. Applying the operator φ^{-1} on the both sides of (Eq. 117), we get:

$$\varphi^{-1}\hat{\mathbf{s}} = \varphi^{-1}(\mathbf{H}^t\mathbf{H} + \eta\mathbf{C}^t\mathbf{C})^{-1}\mathbf{H}^t\mathbf{x} = \varphi^{-1}(\mathbf{H}^t\mathbf{H} + \eta\mathbf{C}^t\mathbf{C})^{-1}\varphi\varphi^{-1}\mathbf{H}^t\varphi\varphi^{-1}\mathbf{x}. \tag{118}$$

The above equation can be easily simplified to the following form:

$$\hat{\mathbf{S}} = (\Lambda_{\mathbf{h}}^*\Lambda_{\mathbf{h}} + \eta\Lambda_{\mathbf{c}}^*\Lambda_{\mathbf{c}})^{-1}\Lambda_{\mathbf{h}}^*\mathbf{X}. \tag{119}$$

This equation can also be expressed in a frequency domain equivalent form as:

$$\hat{S}(k) = \frac{H^*(k)}{|H(k)|^2 + \eta|C(k)|^2} Y(k). \tag{120}$$

6.4 Comparison Study

The LMMSE, the inverse filter, and the regularized speech deconvolution methods have been tested and compared for the case of a lowpass channel with AWGN contamination. Figure 40 shows the original speech signal with its spectrogram. The degraded signal is shown in Fig. 41. The LMMSE, the inverse filter, and the regularized deconvolution results are shown in Figs. 42–44, respectively. These figures show that the best deconvolution results are obtained from the regularized deconvolution method.

The enhancement and deconvolution methods have been applied to the degraded signal at different SNR values, and the results of this comparison are given in

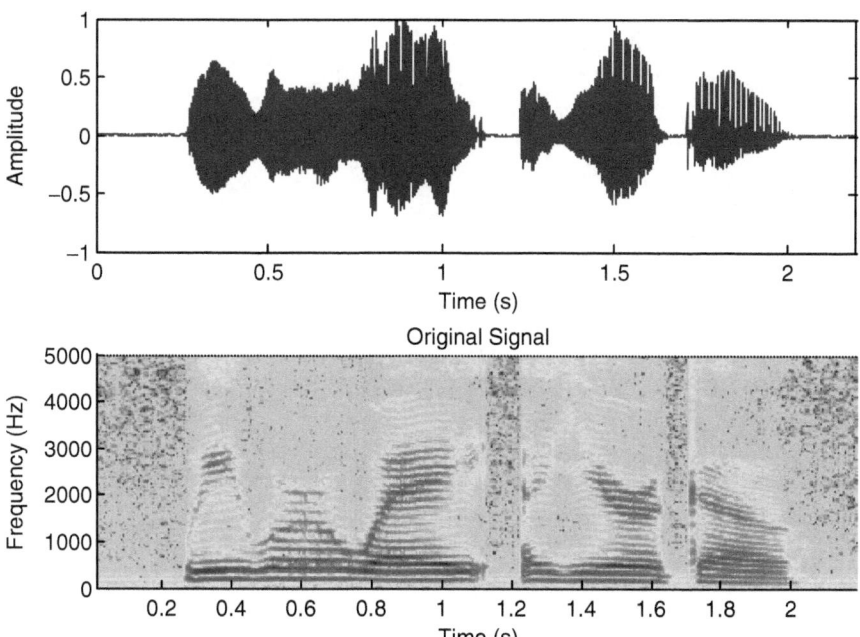

Fig. 40 Time domain waveform and spectrogram of a clean speech signal

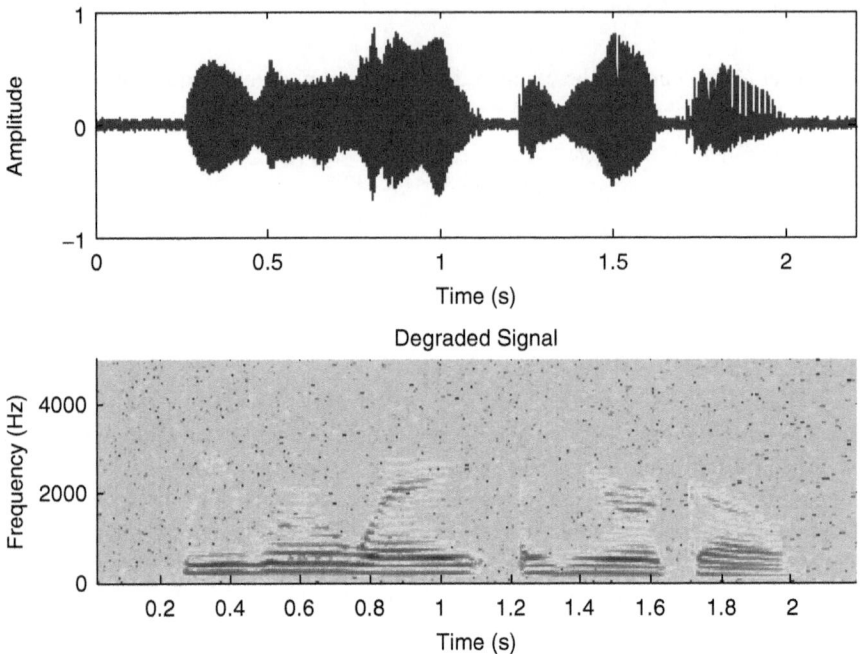

Fig. 41 Time domain waveform and spectrogram of the degraded signal with a lowpass channel effect and AWGN at an $\text{SNR}_{\text{original}} = 20$ dB

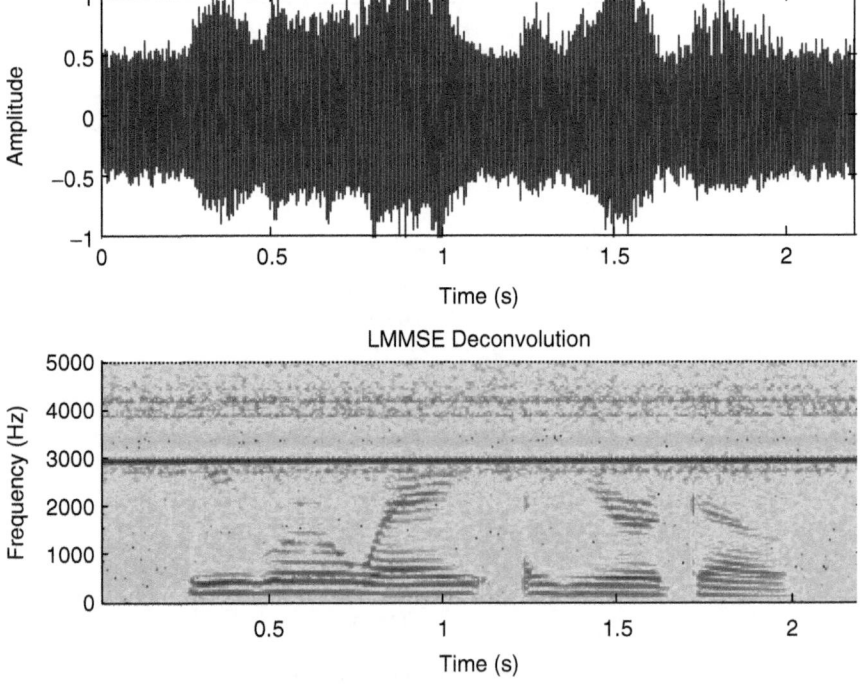

Fig. 42 Time domain waveform and spectrogram of the enhanced signal using the LMMSE deconvolution method, $\text{SNR} = 4.7827$ dB, $\text{SNRseg} = 4.7430$ dB, $\text{LLR} = 0.3797$, $\text{SD} = 8.5461$ dB

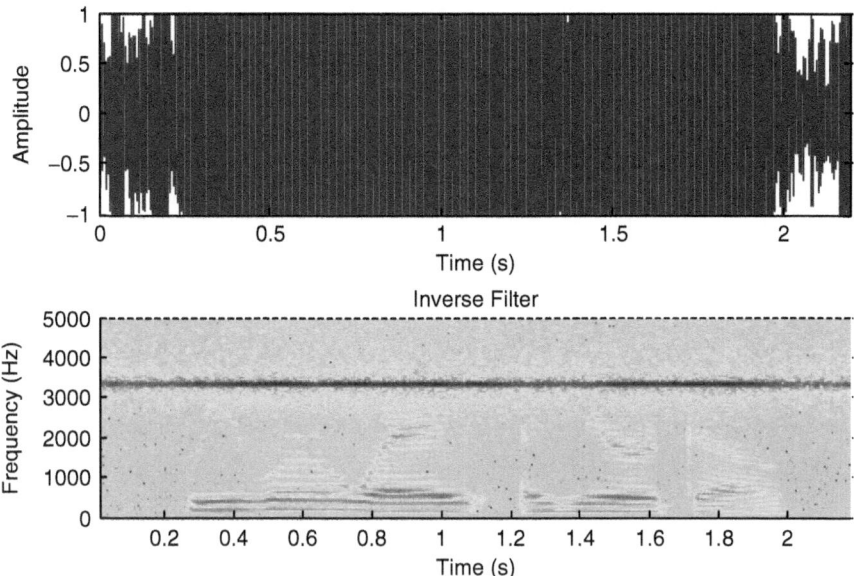

Fig. 43 Time domain waveform and spectrogram of the enhanced signal using the inverse filter deconvolution method, SNR $= 0.0091$ dB, SNRseg $= 0.0091$ dB, LLR $= 0.9698$, SD $= 27.5117$ dB

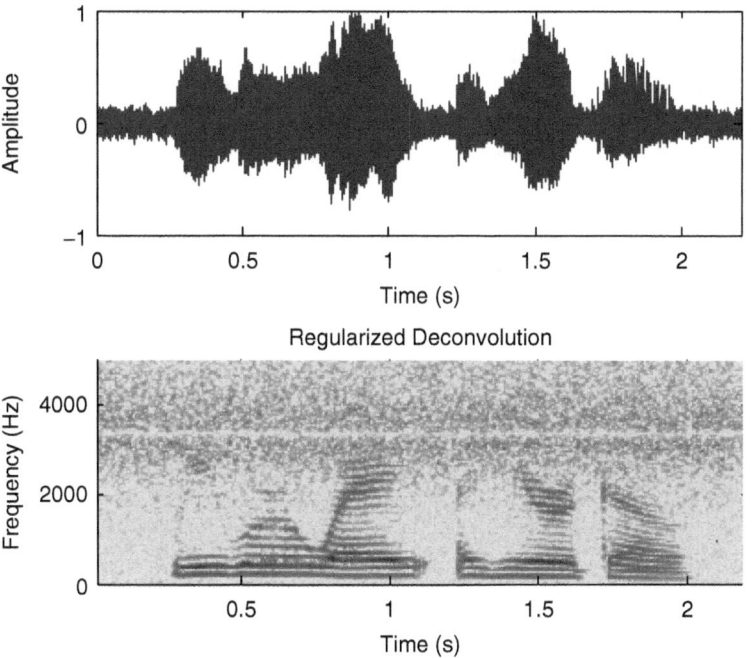

Fig. 44 Time domain waveform and spectrogram of the enhanced signal using the regularized deconvolution method, SNR $= 7.1612$ dB, SNRseg $= 7.0881$ dB, LLR $= 0.2797$, SD $= 7.5155$ dB

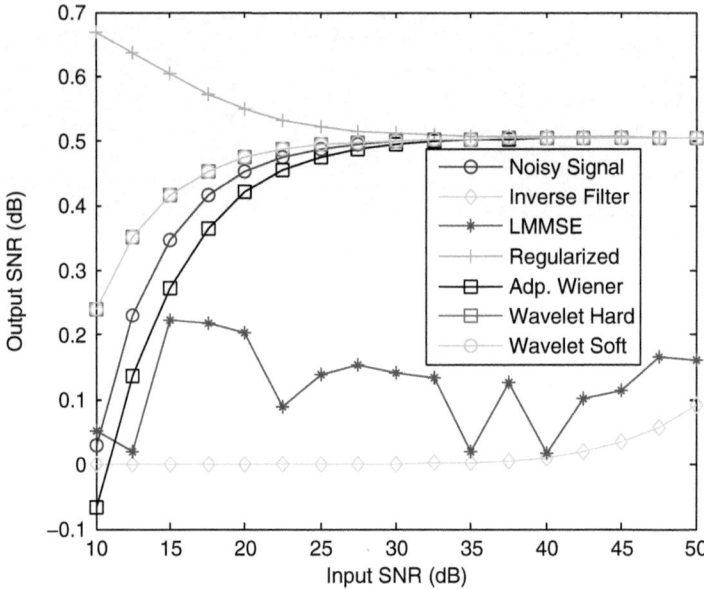

Fig. 45 Output SNR vs. input SNR for enhancement and deconvolution methods

Figs. 45–48. These results are in favor of the regularized deconvolution method as it performs deconvolution under a constraint that preserves the noise at a certain level.

The effect of deconvolution on the performance of speaker identification systems is shown in Fig. 49–51. These figures reveal that regularized deconvolution achieves the best identification scores as compared to the inverse filter and the LMMSE deconvolution methods.

7 Speech Watermarking

Watermarking is a growing field of research, because of its importance for several applications, such as information hiding, copyright protection, fingerprinting, and authentication [59–63]. Watermarking can be applied on speech as well as image, and video signals [59–70]. Speech watermarking can be used in remote access speaker identification systems to increase the degree of security by verifying the existence of the watermark in addition to the identification of the speaker. Several speech watermarking algorithms have been proposed in recent years [64–70]. Watermark embedding through a quantization process is one of the popular speech watermarking algorithms due to its simplicity [64]. Another algorithm is based on the spread spectrum technique, and is implemented by adding pseudo-random sequences to the small segments of the audio signal [65].

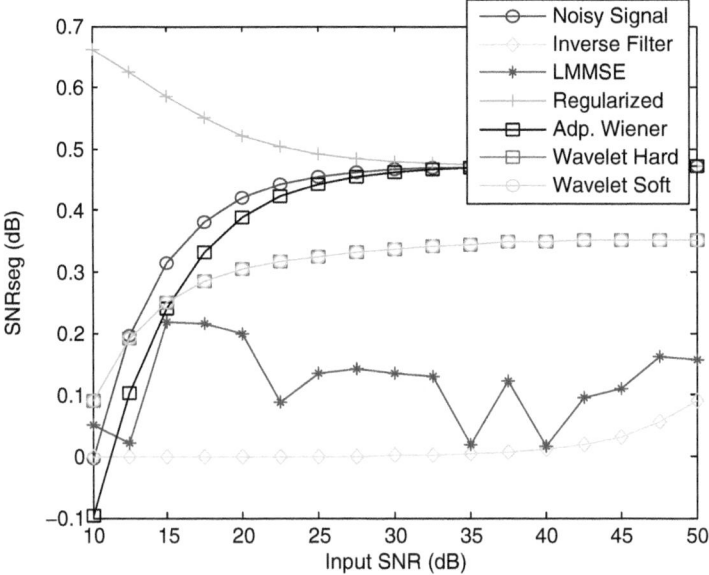

Fig. 46 SNRseg vs. input SNR for enhancement and deconvolution methods

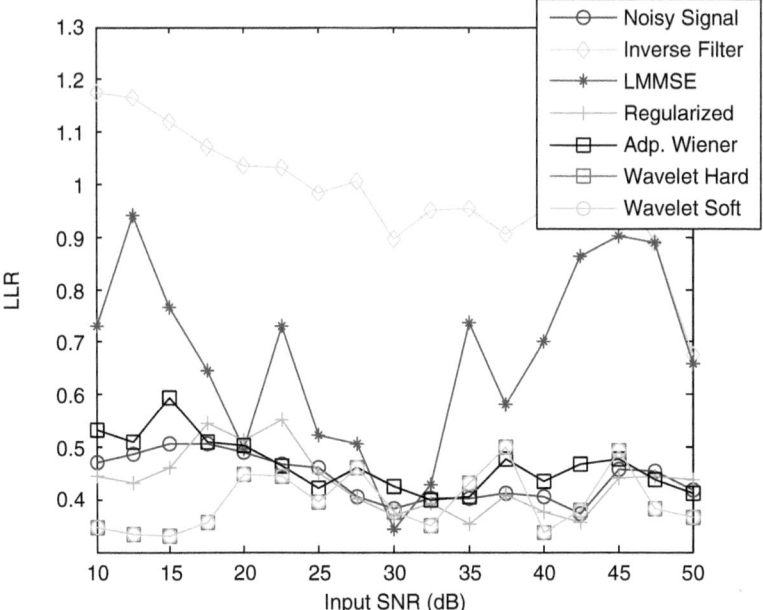

Fig. 47 LLR vs. input SNR for enhancement and deconvolution methods

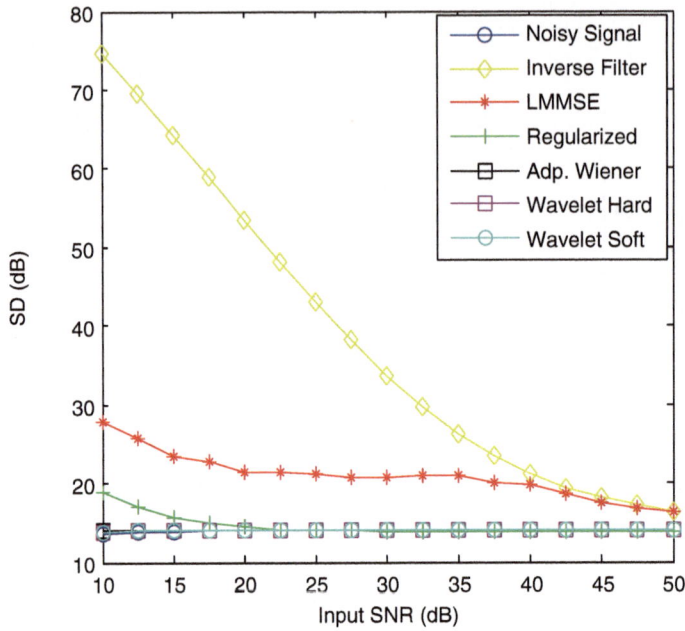

Fig. 48 SD vs. input SNR for enhancement and deconvolution methods

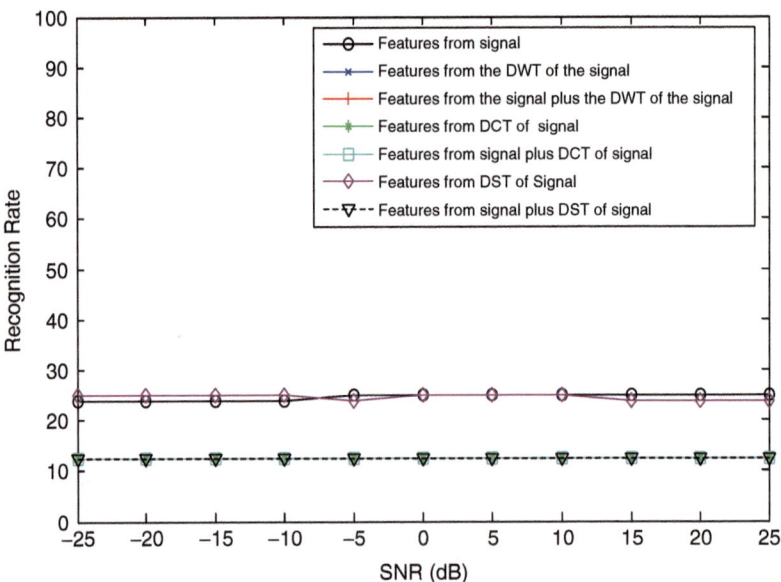

Fig. 49 Recognition rate vs. SNR for the different feature extraction methods with inverse filter deconvolution

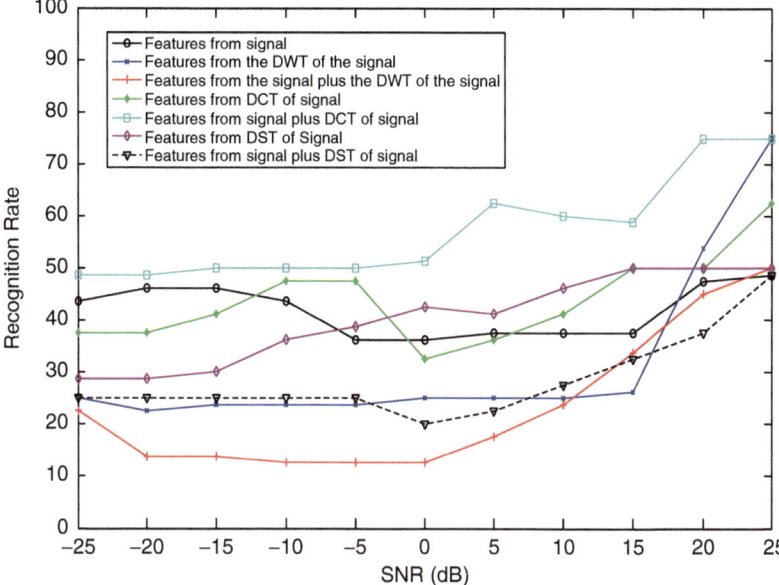

Fig. 50 Recognition rate vs. SNR for the different feature extraction methods with LMMSE deconvolution

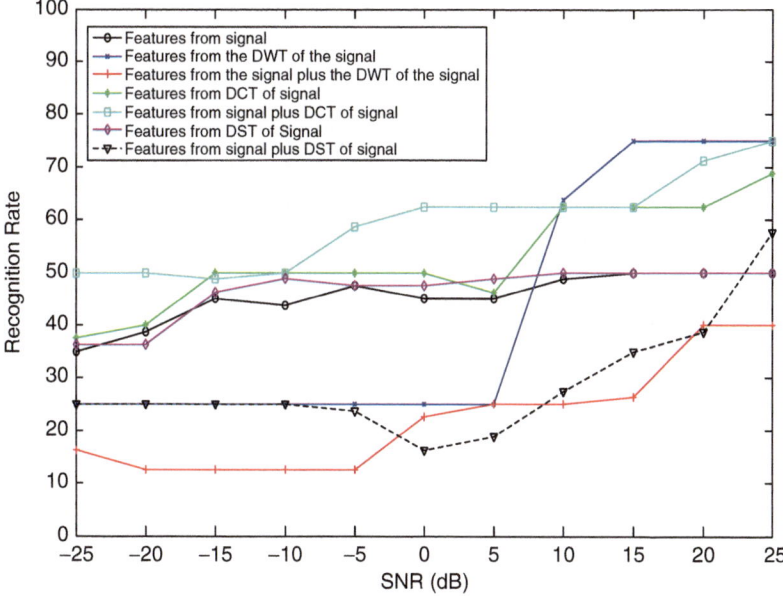

Fig. 51 Recognition rate vs. SNR for the different feature extraction methods with regularized deconvolution

The singular value decomposition (SVD) mathematical technique has also been utilized for speech watermarking in time and transform domains [70, 71]. Concentration in this book will be on the SVD speech watermarking algorithm, because of its ability to embed images in audio signals. With this algorithm, encrypted watermarks can be embedded in speech signals to increase the level of security. The first level of security is the encryption, and the second one is the watermarking. Chaotic encryption is the most appropriate candidate for watermark encryption; because it is a permutation-based encryption algorithm that tolerates channel degradations more efficiently than diffusion-based algorithms [72].

Liu and Tan [73] proposed a watermarking algorithm based on the SVD technique. The main advantage of this algorithm is the robustness against attacks [73–75], because the SVD technique provides an elegant way for extracting algebraic features from a 2D matrix. The singular values (SVs) of a matrix have a good stability. When a small perturbation affects the matrix, no large variations in its SVs occur [73]. Using this property of the SVs of a 2D matrix, the watermark can be embedded in that matrix without a large variation.

7.1   Singular Value Decomposition

The SVD decomposes a matrix $\mathbf{A}$ into three matrices $\mathbf{U}$, $\mathbf{S}$, and $\mathbf{V}$ as follows [76]:

$$\mathbf{A} = \mathbf{U}\mathbf{S}\mathbf{V}^{\mathrm{T}}, \tag{121}$$

where $\mathbf{U}$ and $\mathbf{V}$ are orthogonal matrices such that $\mathbf{U}^{\mathrm{T}}\mathbf{U} = \mathbf{I}$, and $\mathbf{V}^{\mathrm{T}}\mathbf{V} = \mathbf{I}$. $\mathbf{S} = \mathrm{diag}(\sigma_1, \ldots, \sigma_P)$, where $\sigma_1 \geq \sigma_2 \geq, \ldots, \sigma_P \geq 0$ are the SVs of $\mathbf{A}$. The columns of $\mathbf{U}$ are called the left singular vectors of $\mathbf{A}$, and the columns of $\mathbf{V}$ are called the right singular vectors of $\mathbf{A}$.

The properties of the SVD transformation are summarized as follows [76]:

1. The SVs are the square roots of the Eigenvalues.
2. When there is a little disturbance in $\mathbf{A}$, the variations in its SVs are not greater than its largest SV.
3. If the SVs of $\mathbf{A}$ are $\sigma_1, \sigma_2, \ldots, \sigma_P$, the SVs of $\alpha \mathbf{A}$ are $\sigma_1^*, \sigma_2^*, \ldots, \sigma_P^*$, such that:

$$(\sigma_1^*, \sigma_2^*, \ldots, \sigma_P^*) = |\alpha|(\sigma_1, \sigma_2, \ldots, \sigma_P). \tag{122}$$

4. If $\mathbf{P}$ is a unitary and rotating matrix, the SVs of $\mathbf{P}\mathbf{A}$ (rotated matrix) are the same as those of $\mathbf{A}$.
5. The original matrix $\mathbf{A}$ and its shifted versions have the same SVs.
6. Both $\mathbf{A}$ and $\mathbf{A}^{\mathrm{T}}$ have the same SVs.

The above-mentioned properties of the SVD transformation are very much desirable in watermarking. When the watermarked signal with the SVD technique undergoes attacks, the watermark can be retrieved effectively from the attacked watermarked signal.

7.2 The SVD Speech Watermarking Algorithm

This algorithm allows embedding images in audio signals. These images can be extracted at the receiver side. The steps of the embedding algorithm are summarized as follows [71]:

1. The audio signal is either used in time domain or transformed to an appropriate transform domain.
2. The obtained 1D signal is transformed into a 2D matrix ($\mathbf{A}$ matrix).
3. The SVD is performed on the $\mathbf{A}$ matrix as in (Eq. 121).
4. The chaotic encrypted watermark ($\mathbf{W}$ matrix) is added to the SVs of the original matrix.

$$\mathbf{D} = \mathbf{S} + K\mathbf{W}, \tag{123}$$

where K is the watermark weight.
5. The SVD is performed on the new modified matrix ($\mathbf{D}$ matrix).

$$\mathbf{D} = \mathbf{U}_w \mathbf{S}_w \mathbf{V}_w^{\mathrm{T}}. \tag{124}$$

6. The watermarked signal in 2D format ($\mathbf{A}_w$ matrix) is obtained by using the modified matrix of SVs ($\mathbf{S}_w$ matrix).

$$\mathbf{A}_w = \mathbf{U}\mathbf{S}_w\mathbf{V}^{\mathrm{T}}. \tag{125}$$

7. The 2D $\mathbf{A}_w$ matrix is transformed again into a 1D signal.
8. If watermarking is performed in a transform domain, an inverse of this transform is performed.

To extract the possibly corrupted watermark from the possibly distorted watermarked audio signal, given $\mathbf{U}_w$, $\mathbf{S}$, $\mathbf{V}_w$ matrices, and the possibly distorted audio signal, the above steps are reversed as follows:

1. If watermarking is performed in a transform domain, this transform is performed. The 1D obtained signal is transformed to a 2D matrix $\mathbf{A}_w^*$. The * refers to the corruption due to attacks.
2. The SVD is performed on the possibly distorted watermarked image ($\mathbf{A}_w^*$ matrix).

$$\mathbf{A}_w^* = \mathbf{U}^* \mathbf{S}_w^* \mathbf{V}^{*T}. \tag{126}$$

3. The matrix that includes the watermark is computed.

$$\mathbf{D}^* = \mathbf{U}_w \mathbf{S}_w^* \mathbf{V}_w^T. \tag{127}$$

4. The possibly corrupted encrypted watermark is obtained.

$$\mathbf{W}^* = (\mathbf{D}^* - \mathbf{S})/K. \tag{128}$$

5. The obtained matrix $\mathbf{W}^*$ is decrypted.
6. The correlation coefficient between the decrypted matrix and the original watermark is estimated. If this coefficient is higher than a certain threshold, the watermark is present.

7.3 Chaotic Encryption

Chaotic encryption of the watermark image can be performed using the chaotic Baker map. In its discretized form, the Baker map is an efficient tool to randomize a square matrix of data. The discretized map can be represented for an $R \times R$ matrix as follows [77–83]:

$$B(r_1, r_2) = \left[\frac{R}{n_i}(r_1 - R_i) + r_2 \bmod \left(\frac{R}{n_i} \right), \frac{n_i}{R} \left(r_2 - r_2 \bmod \left(\frac{R}{n_i} \right) \right) + R_i \right], \tag{129}$$

where $B(r_1, r_2)$ are the new indices of the data item at (r_1, r_2), $R_i \leq r_1 \leq R_i + n_i$, $0 < r_2 < R$, and $R_i = n_1 + n_2 + \cdots + n_i$.

In steps, the chaotic encryption is performed as follows:

1. An $R \times R$ square matrix is divided into R rectangles of width n_i and number of elements R.
2. The elements in each rectangle are rearranged to a row in the permuted rectangle. Rectangles are taken from left to right beginning with upper rectangles then lower ones.
3. Inside each rectangle, the scan begins from the bottom left corner towards upper elements.

Figure 52 shows an example for the chaotic encryption of an 8×8 square matrix (i.e., $R = 8$). The secret key is $S_{\text{key}} = [n_1, n_2, n_3] = [2, 4, 2]$.

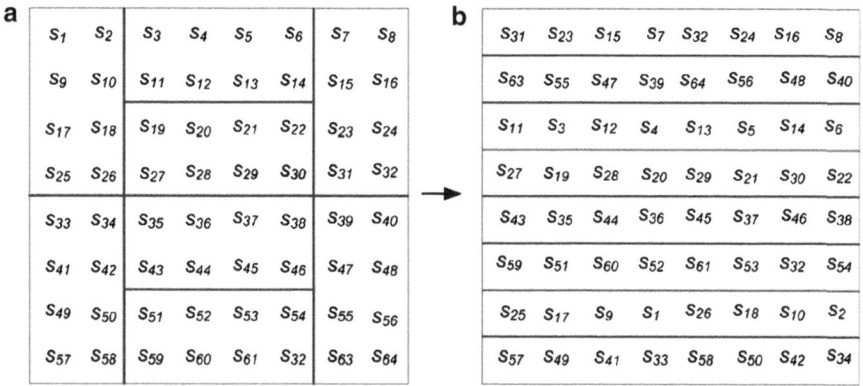

Fig. 52 Chaotic encryption of a square matrix. (**a**) Original square matrix. (**b**) Chaotic encrypted matrix

7.4 The Segment-by-Segment SVD Speech Watermarking Algorithm

If multiple watermarks are added to small speech segments, it is expected that the detectability of the watermark will be enhanced and its robustness against attacks will be increased. Dividing the speech signal into small segments, then embedding the watermark in the SVs of each segment, separately, gives the chance that one or more of the watermarks will survive the attacks, and a higher correlation coefficient in the detection will be obtained.

The original speech signal is divided into nonoverlapping segments. The image watermark is embedded in the SVs (**S** matrix) of each segment after transformation to a small 2D matrix. An SVD is performed on each of these new matrices to get the **S** matrices of the segments. Then, these SV matrices are used to build the watermarked segments.

The steps of the embedding process are summarized as follows [71]:

1. The speech signal is either used in time domain or transformed to a certain transform domain.
2. The obtained signal is divided into nonoverlapping segments and each segment is transformed into a 2D matrix.
3. The SVD is performed on the 2D matrix of each segments ($\mathbf{B}_i$ matrix) to obtain the SVs ($\mathbf{S}_i$ matrix) of each segment, where $i = 1, 2, 3, \ldots, N_s$, and N_s is the number of segments.

$$\mathbf{B}_i = \mathbf{U}_i \mathbf{S}_i \mathbf{V}_i^{\mathrm{T}}. \tag{130}$$

4. The encrypted watermark (**W** matrix) is added to the **S** matrix of each segment.

$$\mathbf{D}_i = \mathbf{S}_i + K\mathbf{W}. \tag{131}$$

5. The SVD is performed on each $\mathbf{D}_i$ matrix to obtain the SVs of each one ($\mathbf{S}_{wi}$ matrix).

$$\mathbf{D}_i = \mathbf{U}_{wi}\mathbf{S}_{wi}\mathbf{V}_{wi}^{\mathrm{T}}. \tag{132}$$

6. The $\mathbf{S}_{wi}$ matrices are used to build the watermarked segments in the time domain.

$$\mathbf{B}_{wi} = \mathbf{U}_i\mathbf{S}_{wi}\mathbf{V}_i^{\mathrm{T}}. \tag{133}$$

7. The watermarked segments are transformed into the 1D format.
8. The watermarked segments are rearranged back into a 1D signal.
9. If watermarking is performed in a transform domain, an inverse of this transform is performed.

Having $\mathbf{U}_{wi}$, $\mathbf{V}_{wi}$, $\mathbf{S}_i$, matrices and the possibly distorted audio signal, we can follow the steps mentioned below to get the possibly corrupted watermark [71].

1. If watermarking is performed in a transform domain, this transform is performed.
2. The possibly corrupted watermarked signal is divided into small segments having the same size used in the embedding process and these segments are transformed into a 2D format.
3. The SVD is performed on each possibly distorted watermarked segment ($\mathbf{B}_{wi}^*$ matrix) to obtain the SVs of each one ($\mathbf{S}_{wi}^*$ matrix).

$$\mathbf{B}_{wi}^* = \mathbf{U}_i^*\mathbf{S}_{wi}^*\mathbf{V}_i^{*\mathrm{T}}. \tag{134}$$

4. The matrices that may contain the watermark are obtained using the $\mathbf{U}_{wi}$, $\mathbf{V}_{wi}$, $\mathbf{S}_{wi}^*$, matrices.

$$\mathbf{D}_i^* = \mathbf{U}_{wi}\mathbf{S}_{wi}^*\mathbf{V}_{wi}^{\mathrm{T}}. \tag{135}$$

5. The possibly corrupted watermark ($\mathbf{W}_i^*$ matrix) is extracted from the $\mathbf{D}_i$ matrices.

$$(\mathbf{D}_i^* - \mathbf{S}_i)/K = \mathbf{W}_i^*. \tag{136}$$

6. The obtained matrix $\mathbf{W}_i^*$ matrices are decrypted.
7. The correlation coefficient between each decrypted matrix $\mathbf{W}_i^*$ and the original watermark is estimated. If at least one of the coefficients is higher than a certain threshold, the watermark is present.

7.5   *Evaluation of SVD Speech Watermarking*

Several experiments have been carried out to test the performance of the SVD speech watermarking algorithm. Time and transform domains have been used for watermark embedding. Both the SVD watermarking algorithm and the segment-by-segment algorithm have been simulated. The CS image is used as a watermark to be embedded in the Handel signal available in Matlab. The original Handel signals with the watermarks used in all experiments are shown in Fig. 53. The correlation coefficient c_r is used to measure the closeness of the obtained watermark to the original watermark.

The effect of the watermark strength K used to add the watermark to the matrix of SVs of the audio signal has been studied. The results of this experiment in the absence of attacks are shown in Figs. 54–58. It is clear from this experiment that in the absence of attacks, watermark embedding in the DFT magnitude or the DST domain achieves the lowest distortion level in the audio signal, but the DST domain

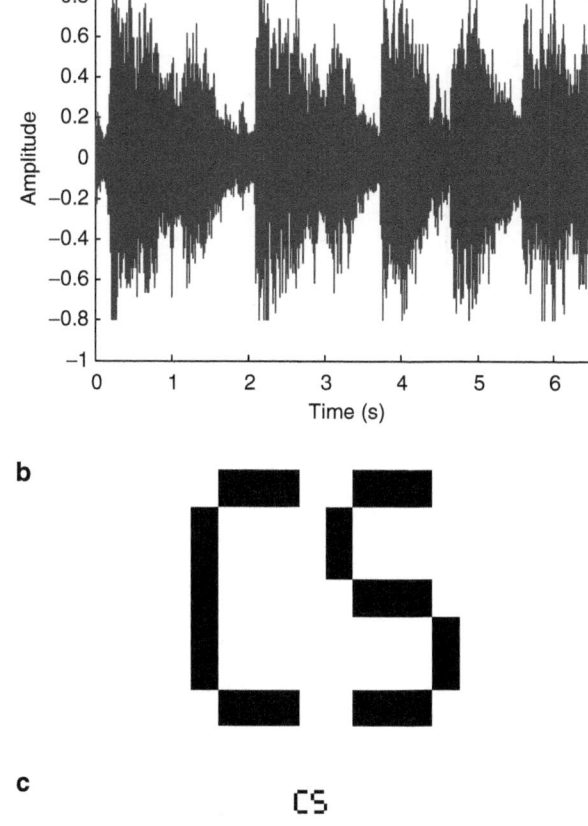

Fig. 53 Original signal and original watermarks. (**a**) Original audio signal. (**b**) Original watermark in the SVD method. (**c**) Original watermark in the segment-by-segment SVD method

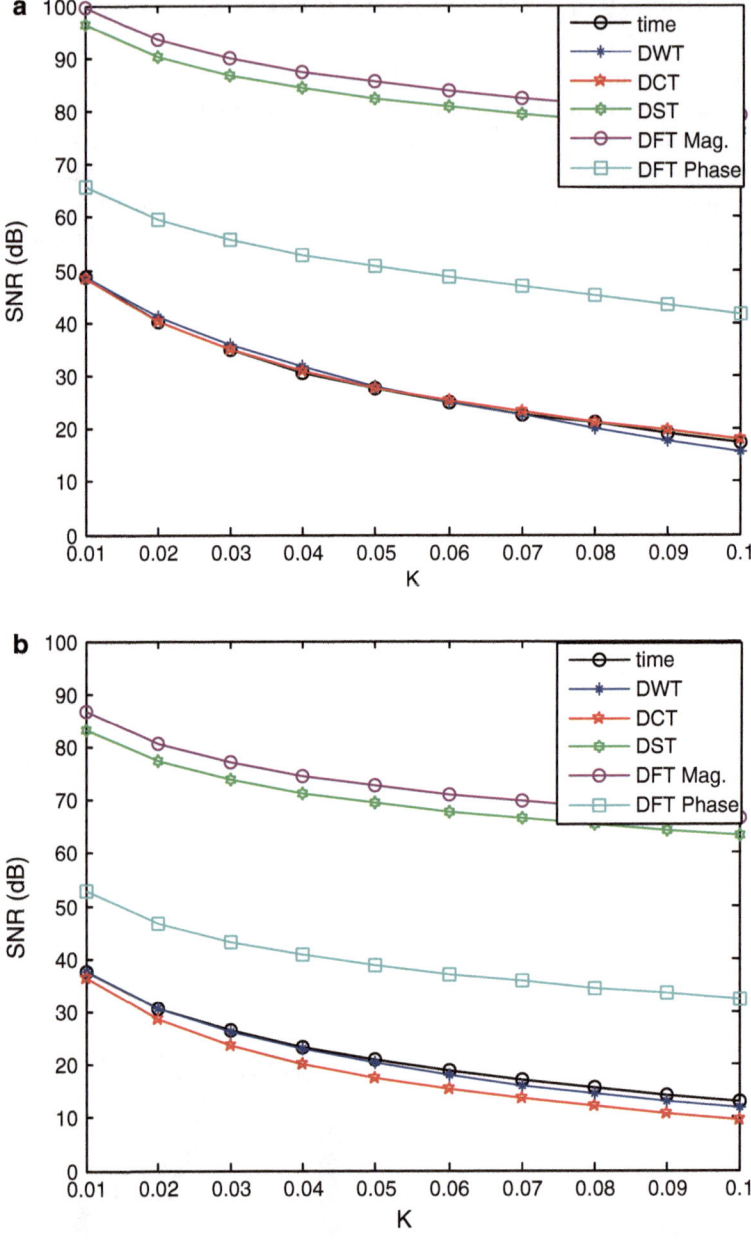

Fig. 54 Variation of the SNR of the watermarked signal with the watermark strength in the absence of any attacks. (**a**) SVD method. (**b**) Segment-by-segment SVD method

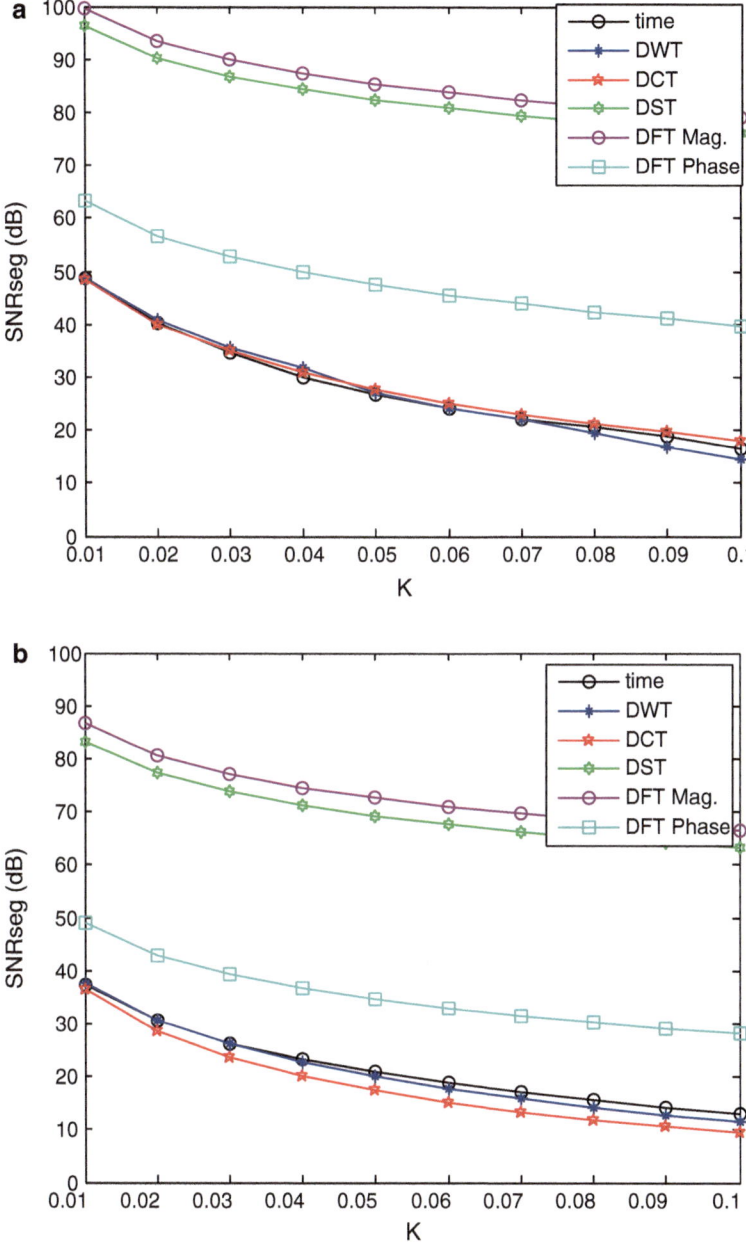

Fig. 55 Variation of the SNRseg of the watermarked signal with the watermark strength in the absence of any attacks. (**a**) SVD method. (**b**) Segment-by-segment SVD method

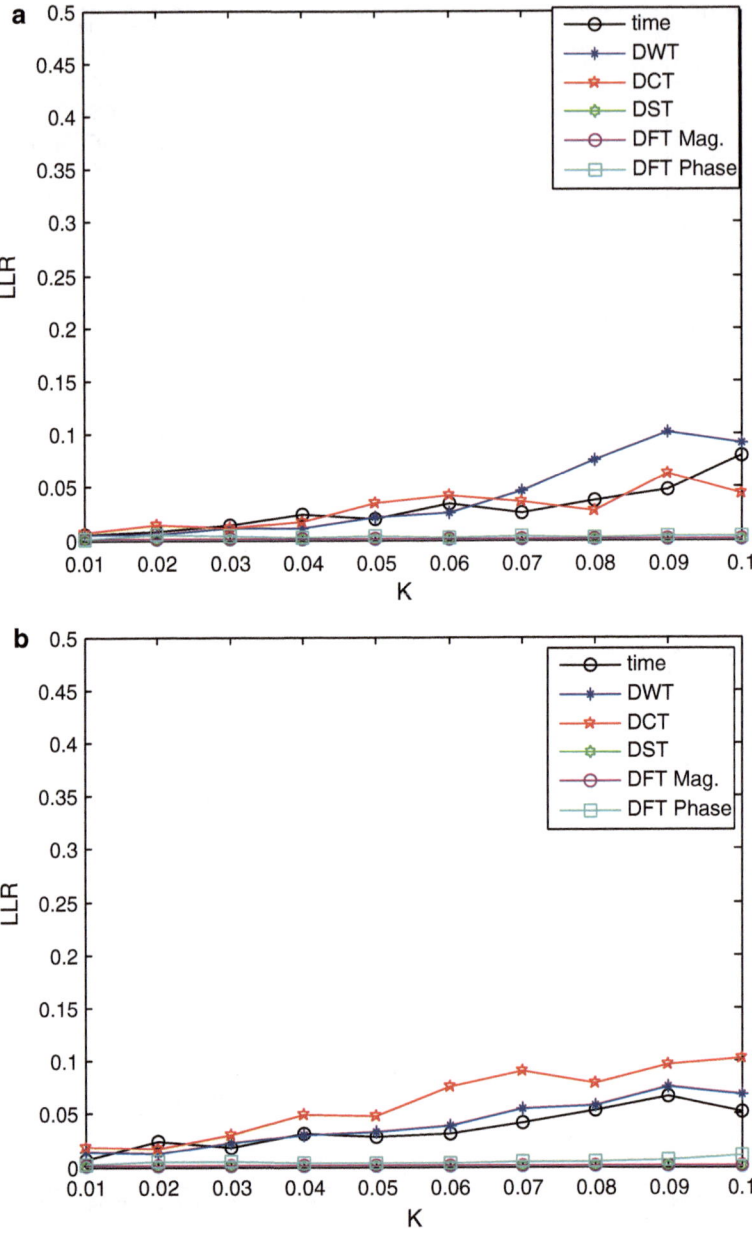

Fig. 56 Variation of the LLR of the watermarked signal with the watermark strength in the absence of any attacks. (**a**) SVD method. (**b**) Segment-by-segment SVD method

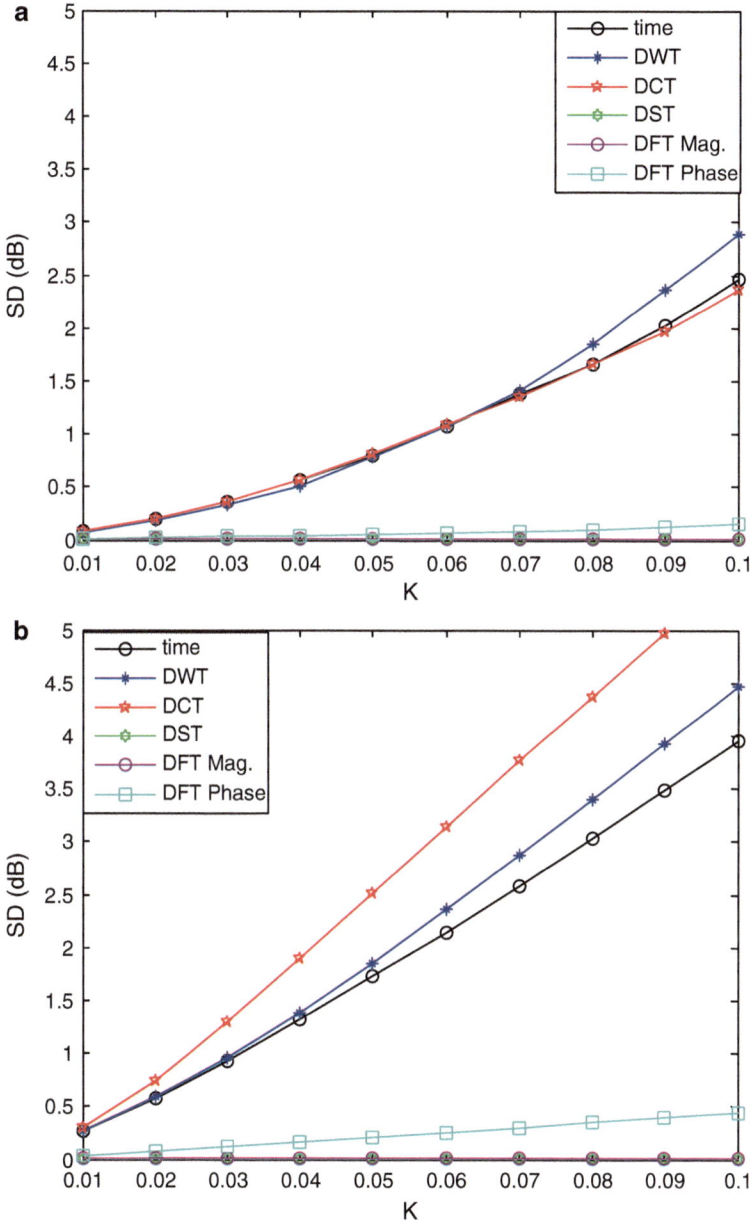

Fig. 57 Variation of the SD of the watermarked signal with the watermark strength in the absence of any attacks. (**a**) SVD method. (**b**) Segment-by-segment SVD method

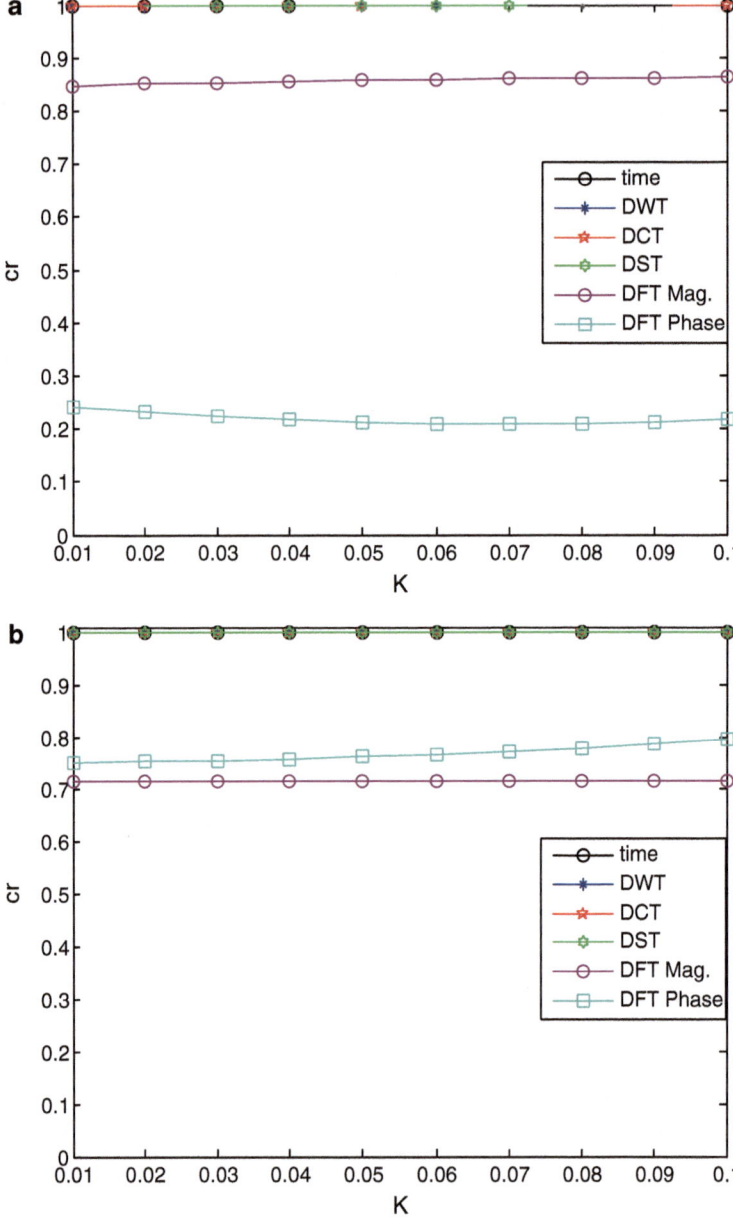

Fig. 58 Variation of the correlation coefficient c_r with the watermark strength in the absence of any attacks. (**a**) SVD method. (**b**) Segment-by-segment SVD method

is preferred to the DFT magnitude in the detection process. It is also clear that segment-by-segment speech watermarking causes more distortion in the audio signal, but achieves more success in the detection in the presence of attacks.

The robustness of both the SVD speech watermarking method and the segment-by-segment SVD method has been studied in the presence of an AWGN attack. Figure 59 shows that watermark embedding in the DWT domain, the DCT domain, or the time domain achieves the highest detection correlation coefficient at low SNR values. From Fig. 57, it is clear that watermark embedding in the time domain achieves the smallest distortion as compared to the DCT domain and the DWT domain, especially for the segment-by-segment SVD method. It is also clear that the segment-by-segment SVD method increases the correlation coefficient of approximately all cases of watermarking.

The robustness of both the SVD speech watermarking method and the segment-by-segment SVD method has been studied in the presence of a lowpass filtering attack. A third order Butterworth filter has been used in this attack. Figure 60 shows that watermark embedding in the time domain achieves the highest detection correlation coefficient for the segment-by-segment SVD method and a sufficiently high correlation coefficient values for the SVD method. It is also clear that the segment-by-segment SVD method increases the correlation coefficient of approximately all cases of time and transforms domain watermarking with the filtering attack, which is a severe case.

The robustness of both the SVD speech watermarking method and the segment-by-segment SVD method has been studied in the presence of a wavelet compression attack. The results of this experiment are given in Fig. 61. Although, watermark embedding in the time domain is not the best case in correlation coefficient values for this attack, the time domain can be chosen as the most appropriate domain for watermark embedding due to the lowest SD, the sufficiently high values of the detection correlation coefficient, and the ability to survive attacks.

The chaotic Baker map has been used to encrypt the watermark image as shown in Fig. 62. The SVD speech watermarking embedding and extraction processes have been performed with an encrypted watermark in the absence of attacks, and the results are shown in Figs. 63 and 64. These figures reveal that the SVD speech watermarking does not degrade the quality of the watermarked audio signal. From the correlation coefficient value between the extracted watermark and the original one, we notice that the watermark is perfectly reconstructed in the absence of attacks.

Four attacks on the watermarked audio signal have been studied; an AWGN attack, a lowpass filtering attack, a cropping attack and a wavelet compression attack. The extracted watermarks in the presence of these attacks are shown in Fig. 65, and the numerical evaluation metrics for these attacks are tabulated in Table 1. This table shows that the correlation coefficient values for the extracted watermarks get lower, but the watermarks are still visible after decryption.

The performance of the segment-by-segment SVD speech watermarking method has been tested and compared to embedding the watermark in the signal as whole. A small encrypted watermark of dimensions 16×16 is embedded in all segments

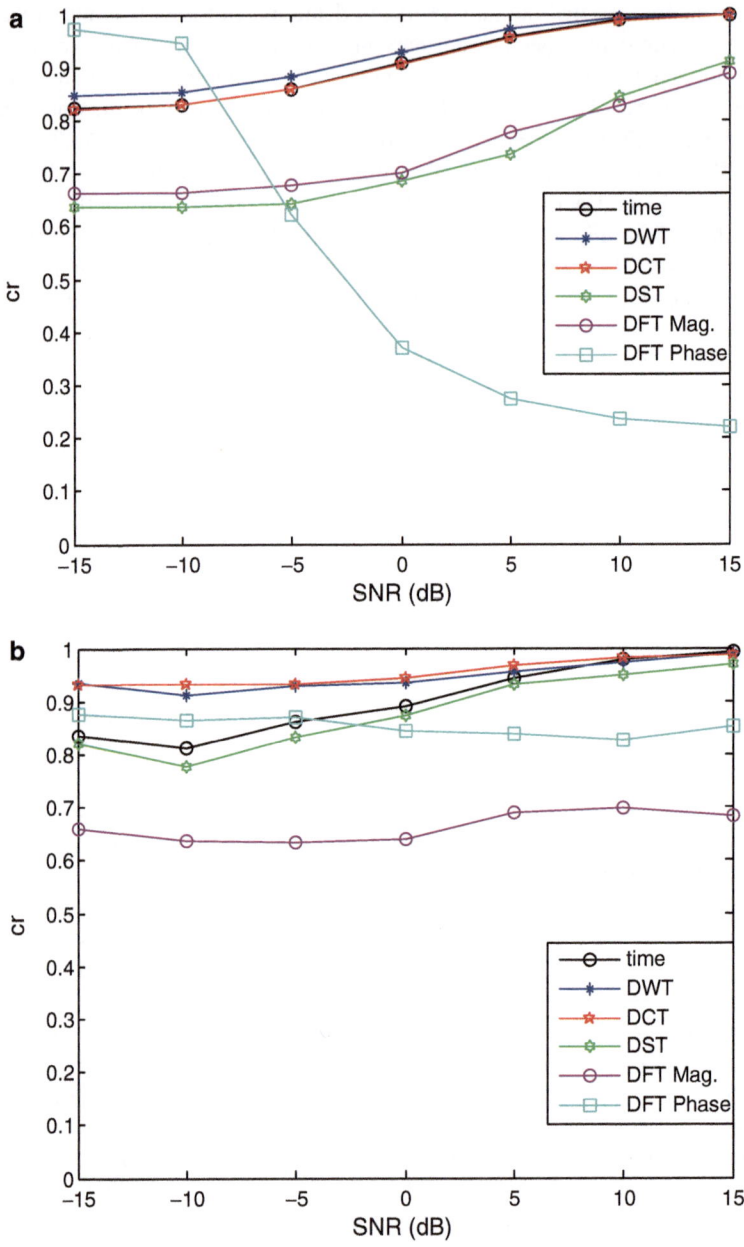

Fig. 59 Variation of the correlation coefficient c_r with the SNR in the presence of AWGN attack. (**a**) SVD method. (**b**) Segment-by-segment SVD method

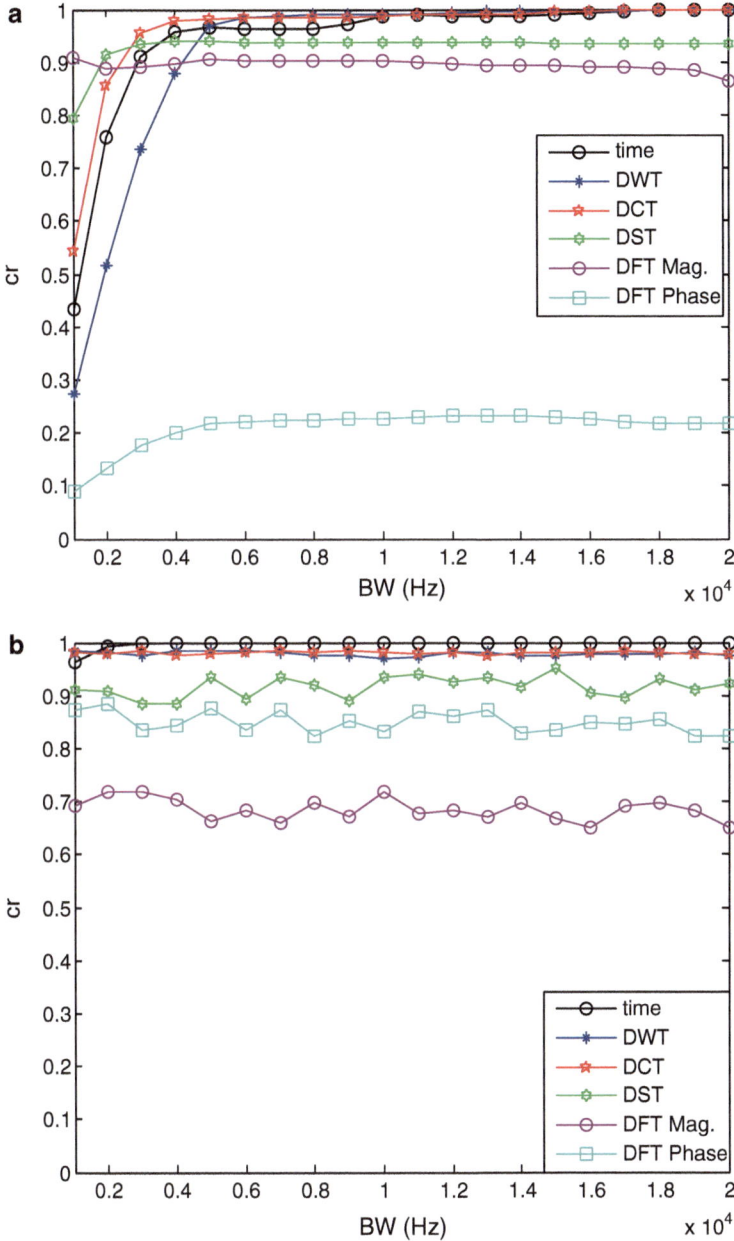

Fig. 60 Variation of the correlation coefficient c_r with the filtering BW in the presence of the filtering attack. (**a**) SVD method. (**b**) Segment-by-segment SVD method

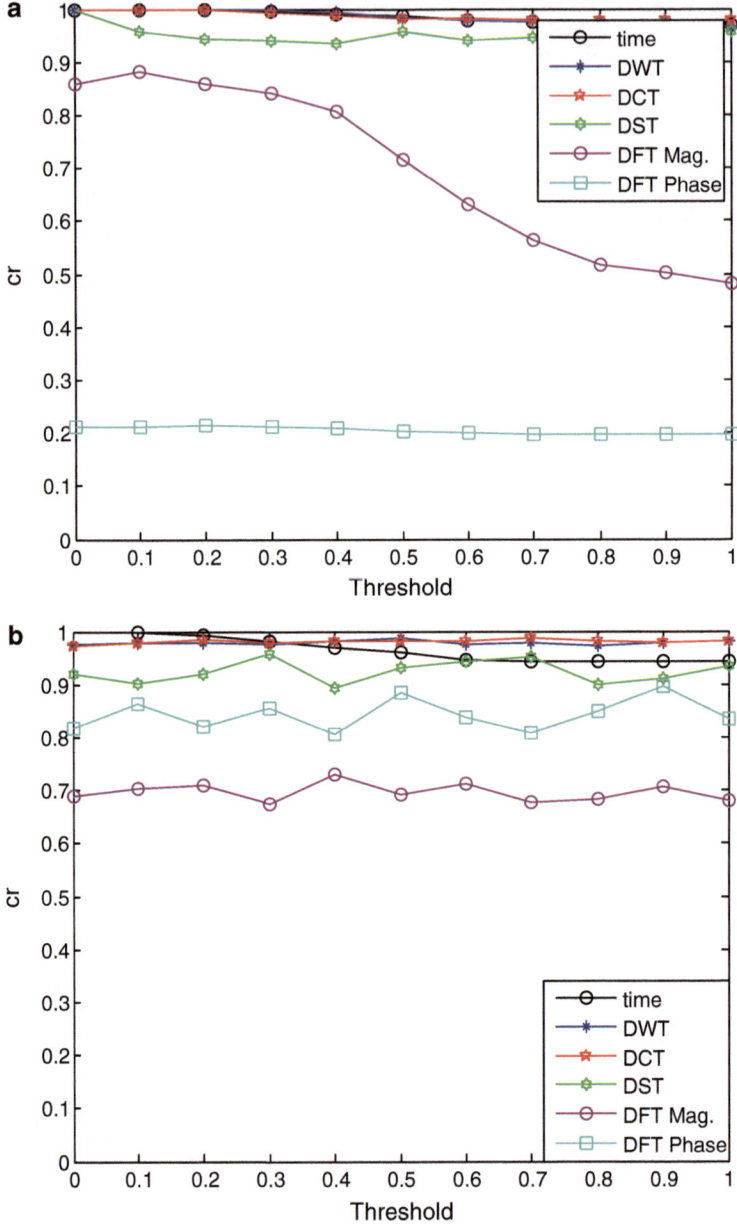

Fig. 61 Variation of the correlation coefficient c_r with the compression threshold in the presence of the wavelet compression attack. (**a**) SVD method. (**b**) Segment-by-segment SVD method

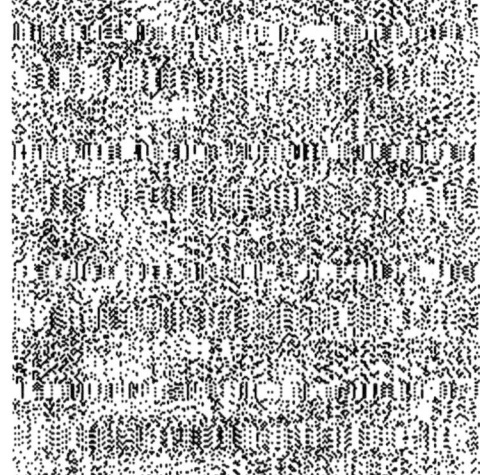

Fig. 62 Chaotic encrypted watermark (CS image) $c_r = 0.0181$

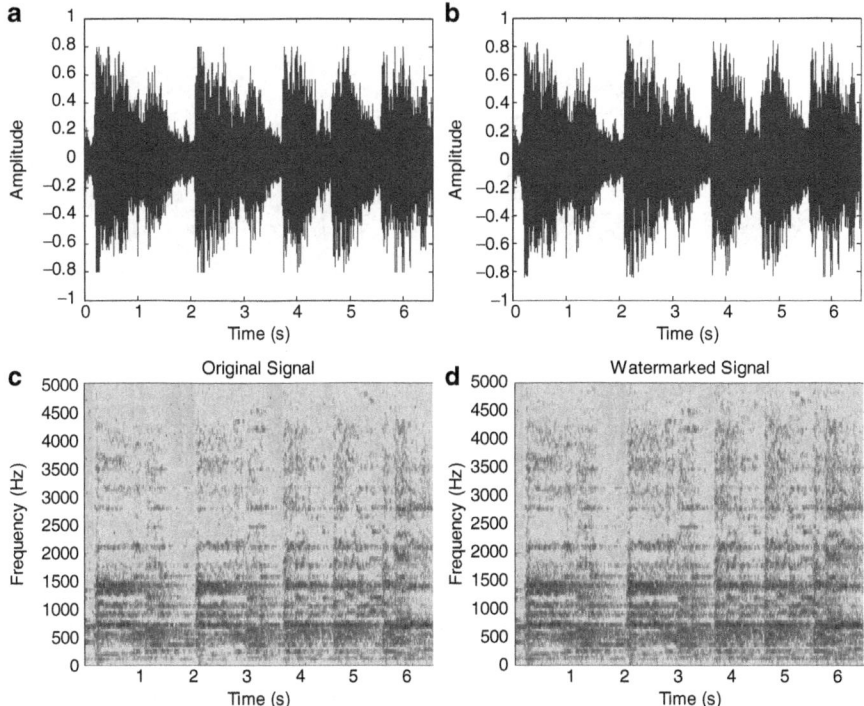

Fig. 63 (**a**) Original audio signal. (**b**) Watermarked audio signal. (**c**) Spectrogram of the original signal. (**d**) Spectrogram of the watermarked signal

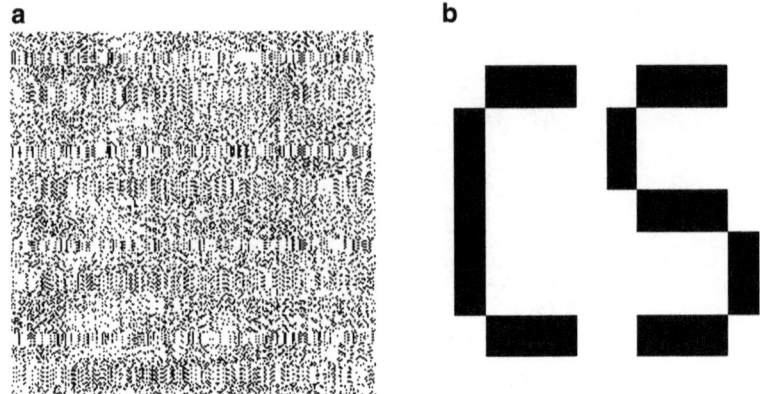

Fig. 64 Extracted watermark without attacks. (**a**) Encrypted watermark $c_r = 0.0181$. (**b**) Decrypted watermark $c_r = 1$

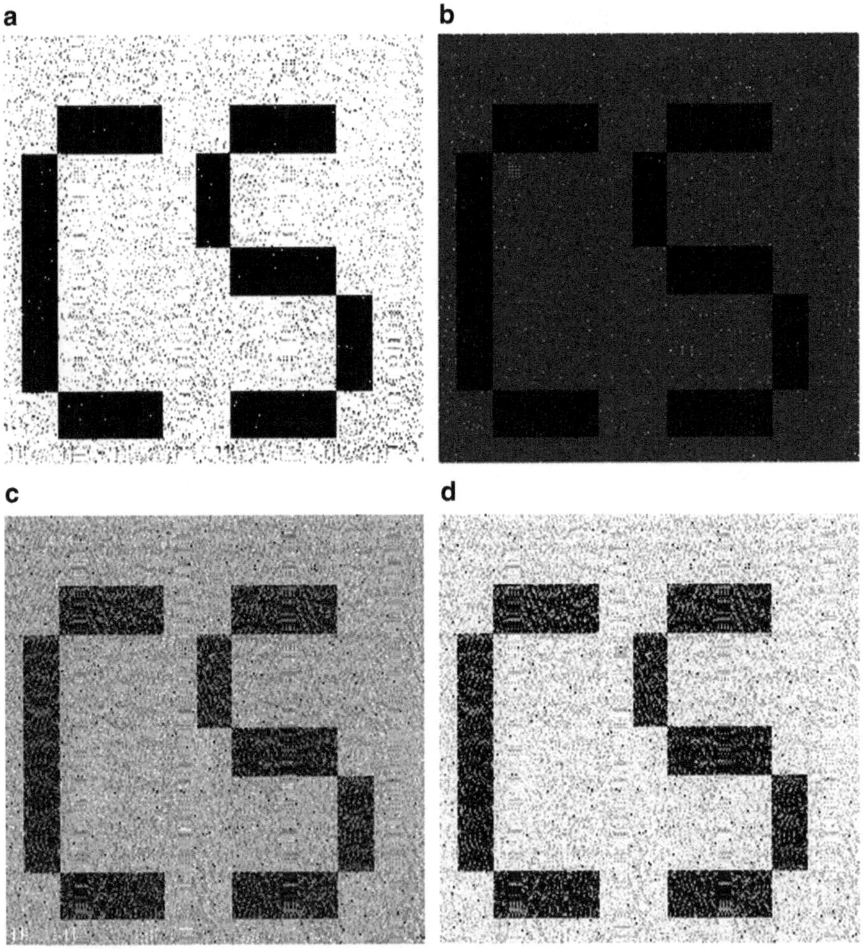

Fig. 65 Extracted watermark in the presence of attacks. (**a**) Noise attack. (**b**) Filtering attack. (**c**) Cropping attack. (**d**) Wavelet compression attack

Table 1 Numerical evaluation metrics for the SVD watermarking method of the audio signal as whole

	No attacks	Noise attack	Filtering attack	Cropping attack	Compression attack
SNR (dB)	27.13	−10.26	1.56	3.03	9.1
SNRseg (dB)	26.31	−10.29	1.56	3	9.04
LLR	0.02	0.34	0.39	0.27	0.12
SD	0.84 dB	−0.01	34.7 dB	11.96 dB	6.07 dB
c_{re}	0.02	26.02 dB	0.0006	−0.006	0.02
c_{rd}	1	0.26	0.02	0.16	0.54

c_{re} is the correlation coefficient between the extracted encrypted watermark and the original watermark. c_{rd} is the correlation coefficient between the extracted decrypted watermark and the original watermark

Fig. 66 The block watermark used for segment-by-segment watermarking. (**a**) Original watermark. (**b**) Chaotic encrypted watermark

of the audio signal. The length of each segment is 256 samples, which is the number of pixels in the small watermark. No overlapping is implemented between segments. The watermark image and its chaotic encrypted version are shown in Fig. 66. The results of the segment-by-segment SVD method in the presence of attacks are shown in Figs. 67 and 68. These figures show that the segment-by-segment method is similar in performance to the SVD method in the absence of attacks. The audio signals are still not deteriorated due to segment-by-segment watermarking. The extracted watermarks for the segment-by-segment method in the presence of attacks are shown in Fig. 69. The numerical evaluation metrics for these results are tabulated in Table 2. From this table, we notice that the correlation coefficient between, at least, one of the extracted watermarks and the original watermark exceeds the corresponding value obtained from the SVD watermarking of the audio signal as a whole. The segment-by-segment watermarking enables PR of the embedded watermark in the presence of the cropping attack.

Figures 70–72 show a comparison between the SVD watermarking method of the audio signal as a whole and the segment-by-segment SVD method in the presence of the AWGN attack, the filtering attack, and the wavelet compression attack, respectively. From the results in these figures, we can conclude that for a low SNR environment, the segment-by-segment method is preferred, because it increases the detection probability of the watermark. For the filtering attack, although the extracted watermark has a low correlation coefficient with the original one, because the filter removes most of the signal details, the segment-by-segment method has a better performance than the SVD watermarking method on the signal as a whole. For the wavelet compression attack, it is clear that as the threshold below which wavelet coefficients are neglected increases, the segment-

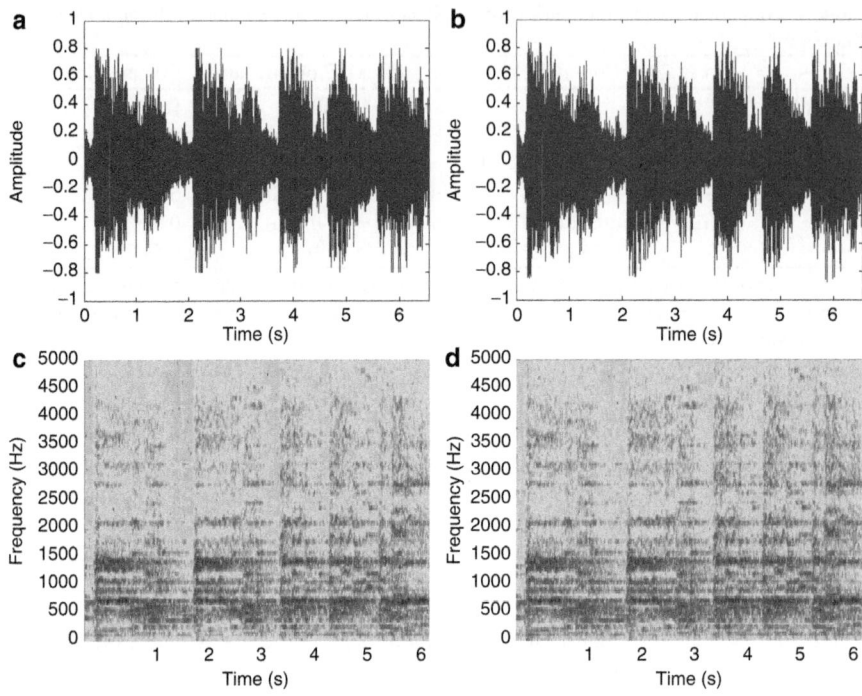

Fig. 67 (**a**) Original audio signal. (**b**) Segment-by-segment SVD watermarked signal. (**c**) Spectrogram of the original signal. (**d**) Spectrogram of the watermarked signal

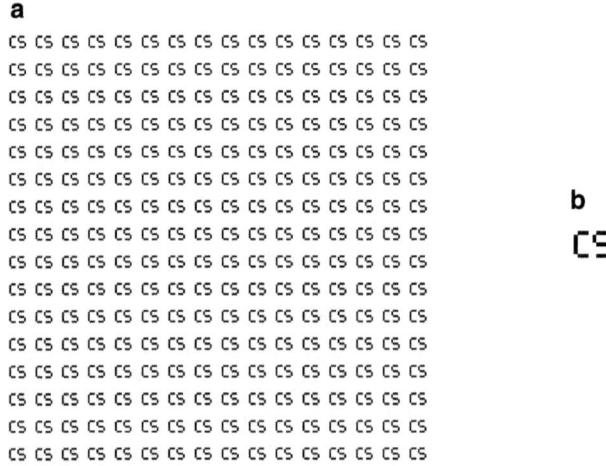

Fig. 68 Extracted watermarks without attacks. (**a**) Watermarks after decryption. (**b**) Decrypted watermark which has $c_{rmax} = 1$

by-segment SVD method achieves a better performance than the SVD method on the audio signal as a whole. This is attributed to the high probability that, at least, one of the several watermarks will not be affected by the compression process.

Speaker identification is usually used as a tool of security. To increase the degree of security, it is recommended to add encrypted watermarks to the speech signals that will be used for speaker identification. If the speaker is verified from the features of his speech and the watermark were found, this can be used as a complicated authentication tool. The effect of watermark embedding on speech features, and hence the performance of the speaker identification system should be considered.

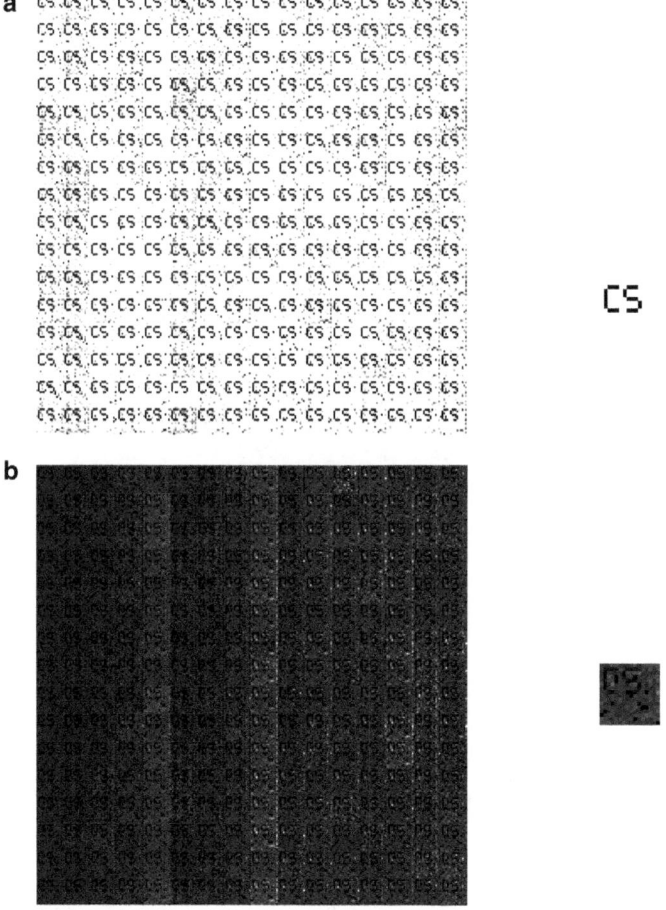

Fig. 69 Extracted watermarks with attacks. (**a**) Noise attack. (**b**) Filtering attack. (**c**) Cropping attack. (**d**) Wavelet compression attack (*Left*: all extracted watermarks and *Right*: watermark which achieves maximum correlation coefficient with the original watermark)

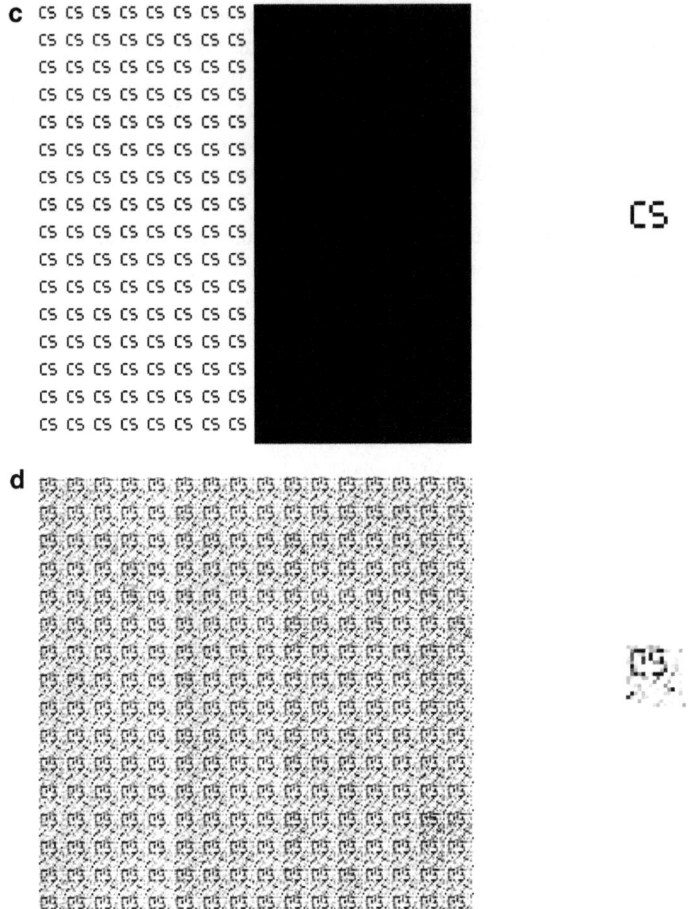

Fig. 69 (continued)

Table 2 Numerical evaluation metrics for the segment-by-segment SVD watermarking method

	No attacks	Noise attack	Filtering attack	Cropping attack	Compression attack
SNR (dB)	21.37	−10.54	1.6	3.01	8.97
SNRseg (dB)	21.29	−10.57	1.6	2.98	8.91
LLR	0.04	0.36	0.39	0.26	0.16
SD	1.6 dB	26.53	34.1 dB	11.74 dB	6.05 dB
c_{rmax}	1	0.34	0.07	1	0.72

c_{rmax} is the correlation coefficient between the extracted watermark, which achieves maximum correlation with original watermark and the original watermark

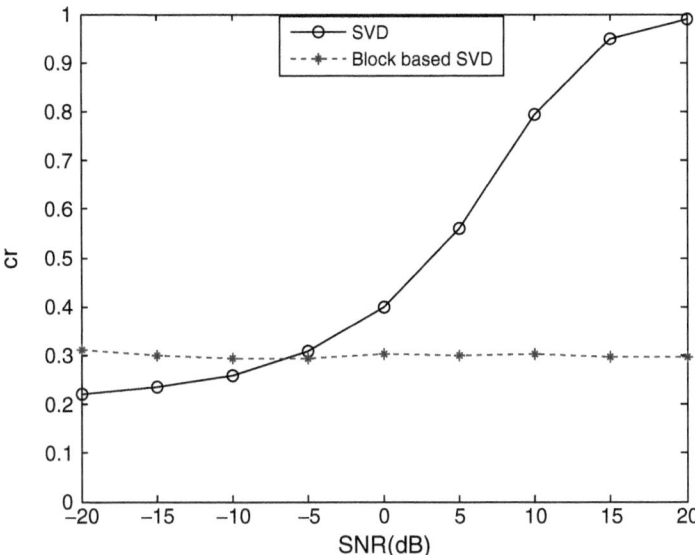

Fig. 70 Correlation coefficient between the extracted watermark and the original watermark vs. the SNR for both the SVD and the segment-by-segment SVD watermarking methods in the presence of AWGN attack

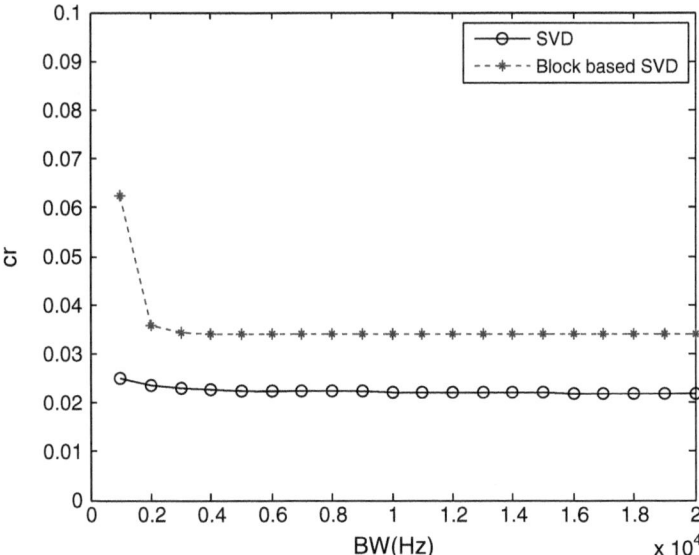

Fig. 71 Correlation coefficient between the extracted watermark and the original watermark vs. the filter BW for both the SVD and the segment-by-segment SVD watermarking methods in the presence of a filtering attack

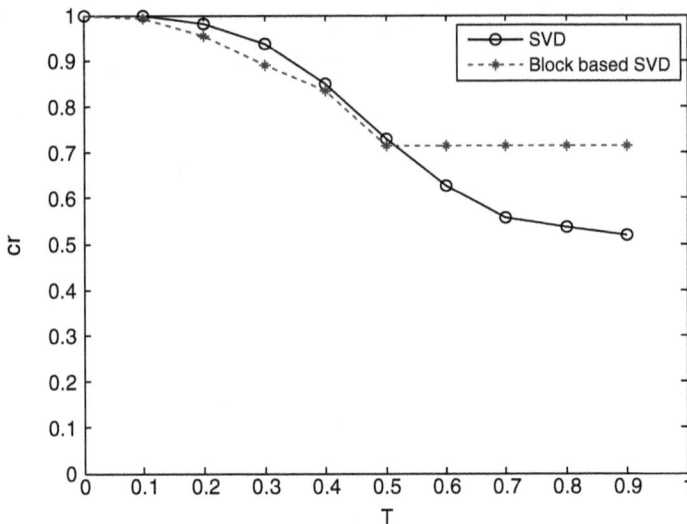

Fig. 72 Correlation coefficient between the extracted watermark and the original watermark vs. the wavelet compression threshold for both the SVD and the segment-by-segment SVD watermarking methods in the presence of a wavelet compression attack

The performance of the speaker identification system has been tested with the SVD watermarking to increase the level of security by using encrypted watermarks. The results of some experiments carried out to test the performance of the speaker identification system with the SVD watermarking method are shown in Figs. 73–83. Some other experiments have also been carried out with the segment-by-segment SVD method, and the results are given in Figs. 84–94. From these results, it is clear that speech watermarking does not degrade the speaker identification system performance. So, it can be used in speaker identification systems to increase security. In [71], it was shown the segment-by-segment watermarking in the time domain achieves the highest detectability of the watermark. So, it is recommended to use the segment-by-segment SVD method with speaker identification systems implementing features extracted from the DCT or the DWT.

8 Speech Encryption

Speech encryption can be used as a tool to prevent eavesdroppers from getting the speech signals that will be used for feature extraction. The main objective of speech encryption is to avoid any unauthorized access to the system of concern by synthesis trials. Speech encryption seeks to perform a completely reversible operation on speech to be totally unintelligible to any unauthorized listener [84]. In speech

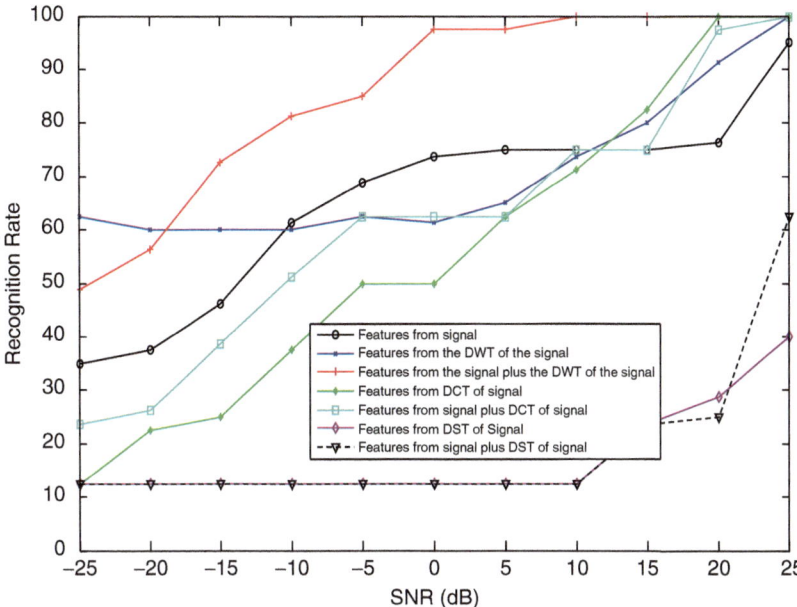

Fig. 73 Recognition rate vs. SNR for the different feature extraction methods with SVD speech watermarking in the time domain and AWGN

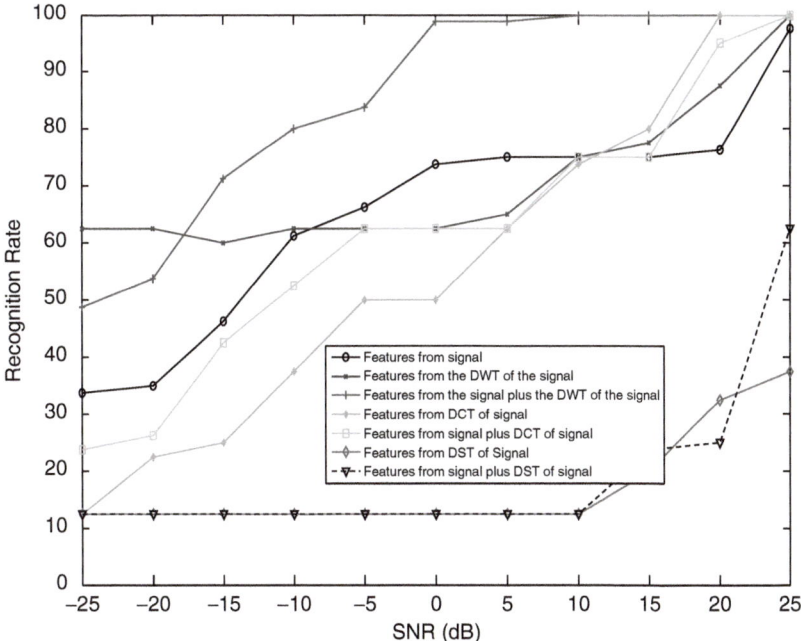

Fig. 74 Recognition rate vs. SNR for the different feature extraction methods with SVD speech watermarking in the time domain and Wiener filtering

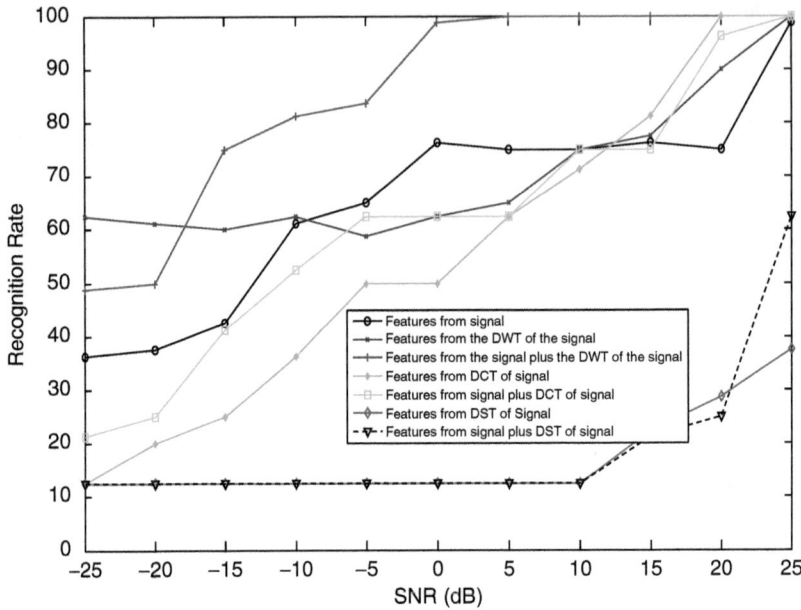

Fig. 75 Recognition rate vs. SNR for the different feature extraction methods with SVD speech watermarking in the time domain and spectral subtraction

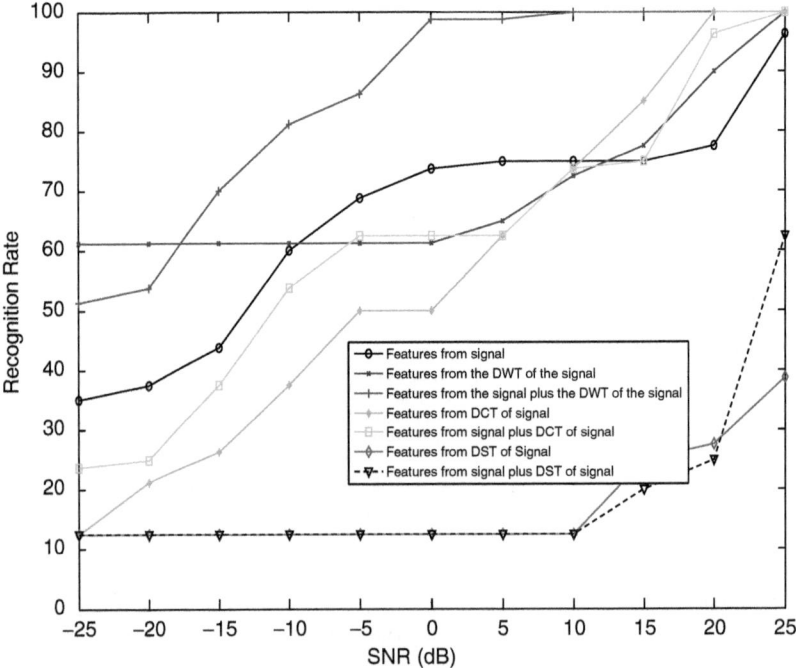

Fig. 76 Recognition rate vs. SNR for the different feature extraction methods with SVD speech watermarking in the time domain and adaptive Wiener filtering

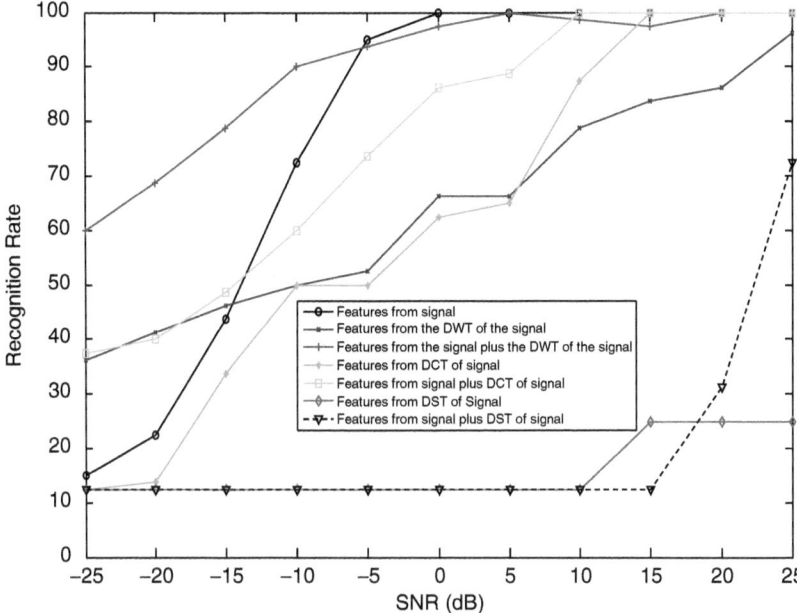

Fig. 77 Recognition rate vs. SNR for the different feature extraction methods with SVD speech watermarking in the time domain and wavelet soft thresholding with 1 level Haar wavelet

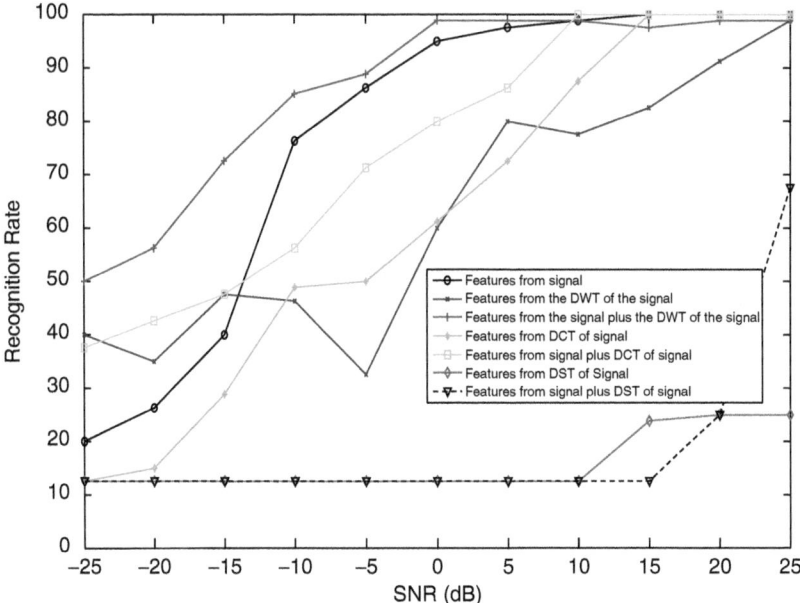

Fig. 78 Recognition rate vs. SNR for the different feature extraction methods with SVD speech watermarking in the time domain and wavelet hard thresholding with 1 level Haar wavelet

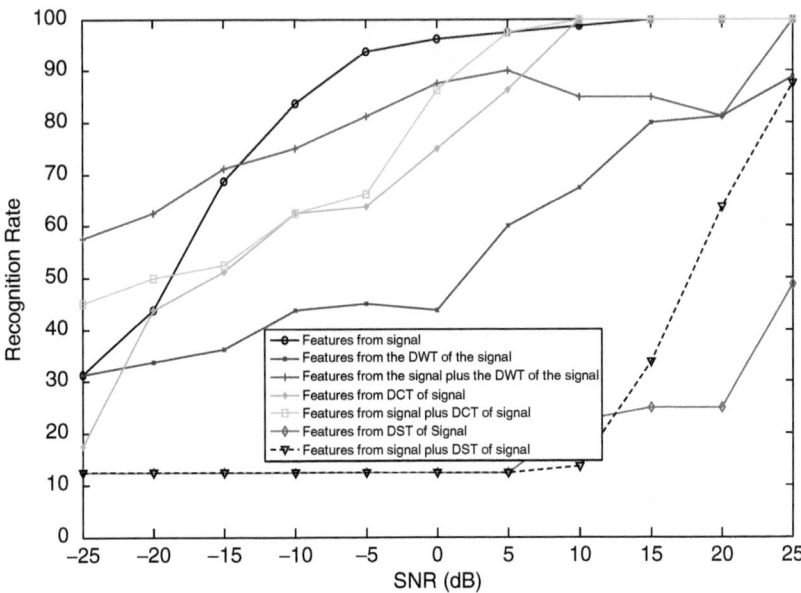

Fig. 79 Recognition rate vs. SNR for the different feature extraction methods with SVD speech watermarking in the time domain and wavelet soft thresholding with 2 levels Haar wavelet

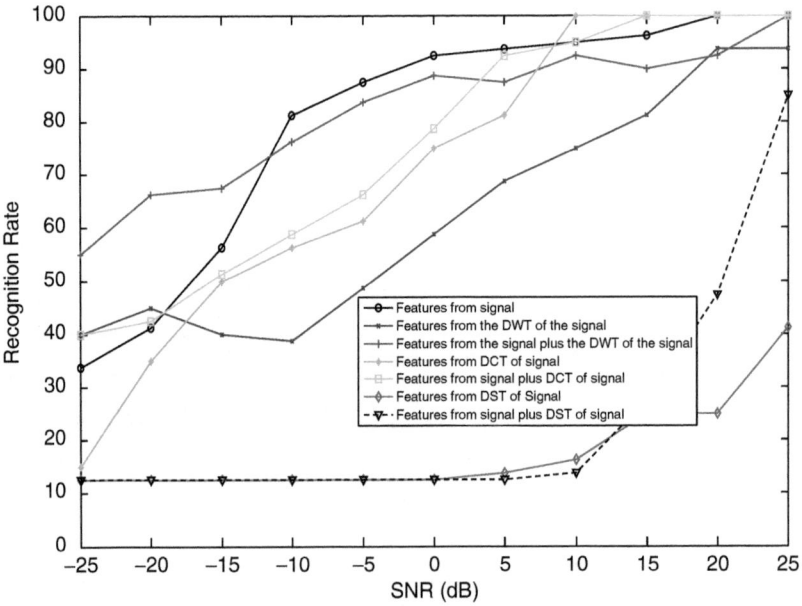

Fig. 80 Recognition rate vs. SNR for the different feature extraction methods with SVD speech watermarking in the time domain and wavelet hard thresholding with 2 levels Haar wavelet

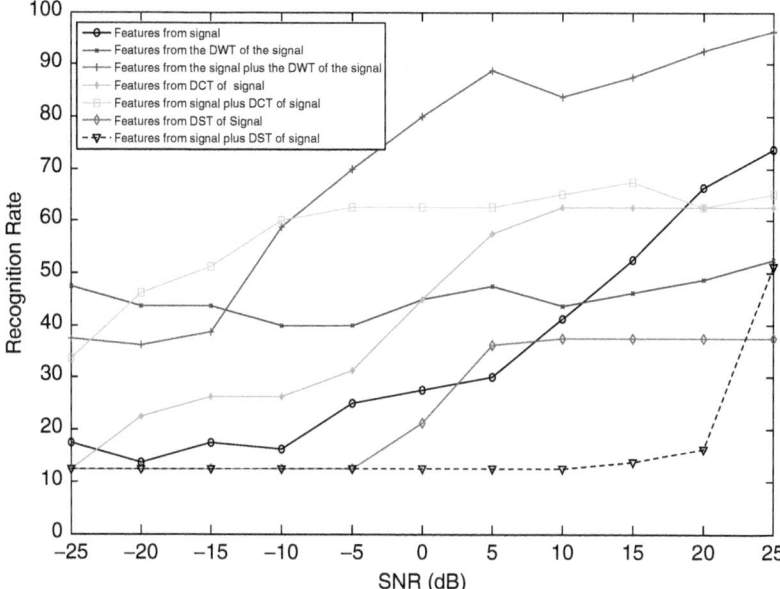

Fig. 81 Recognition rate vs. SNR for the different feature extraction methods with SVD speech watermarking in the time domain and inverse filter deconvolution

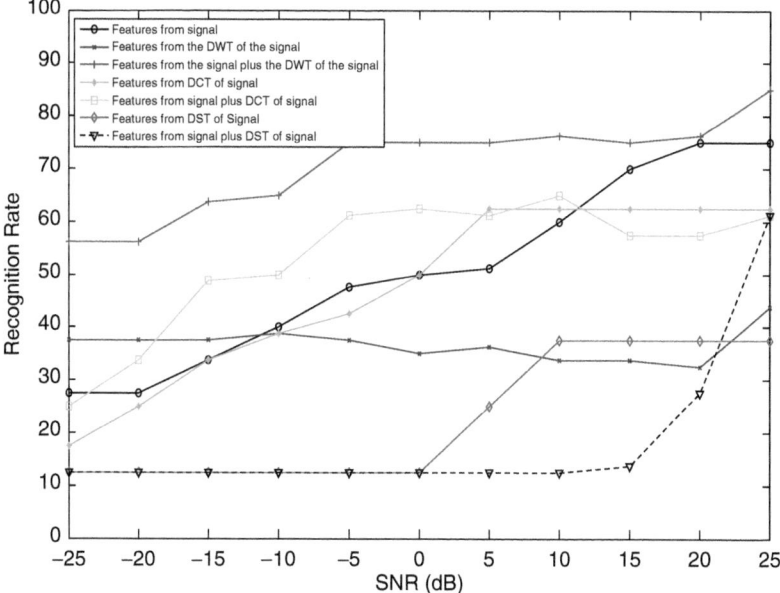

Fig. 82 Recognition rate vs. SNR for the different feature extraction methods with SVD speech watermarking in the time domain and LMMSE deconvolution

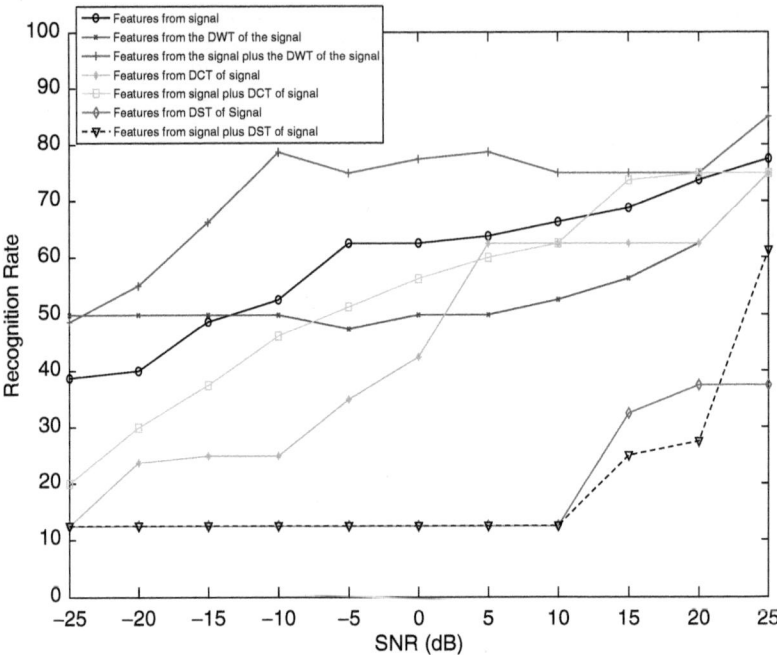

Fig. 83 Recognition rate vs. SNR for the different feature extraction methods with SVD speech watermarking in the time domain and regularized deconvolution

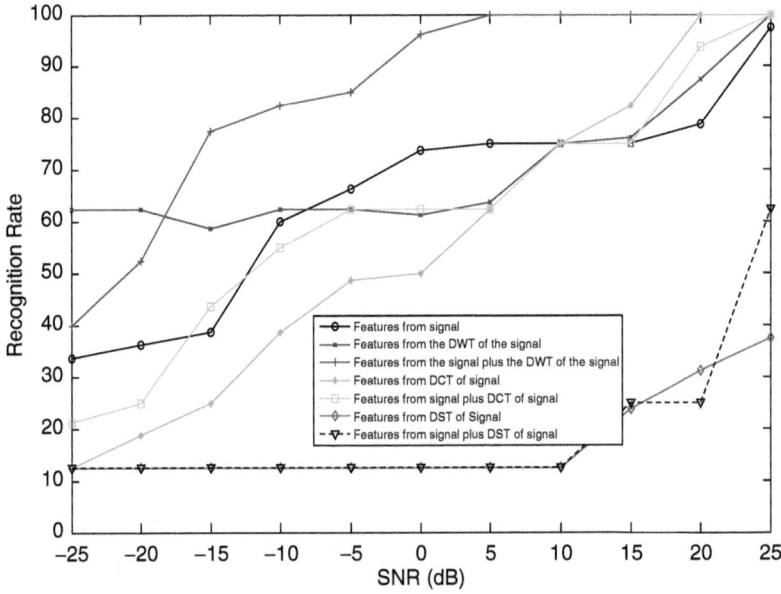

Fig. 84 Recognition rate vs. SNR for the different feature extraction methods with segment-by-segment SVD speech watermarking in the time domain and AWGN

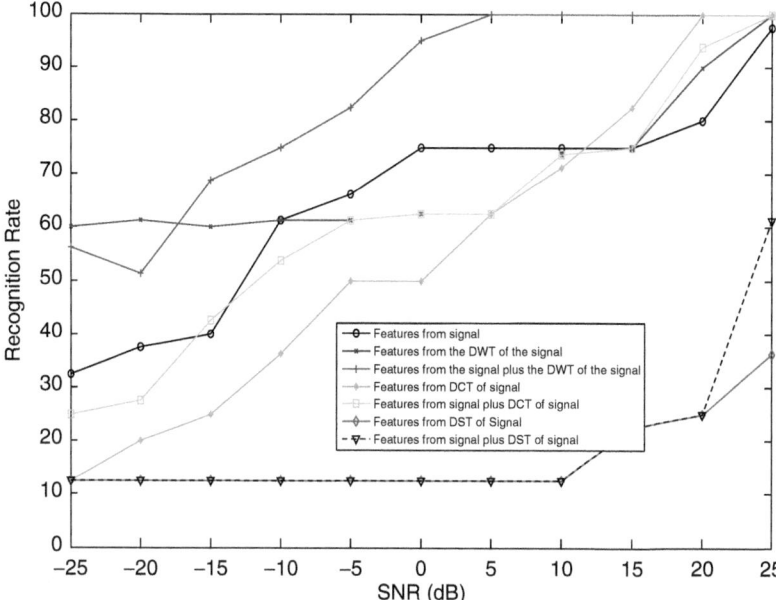

Fig. 85 Recognition rate vs. SNR for the different feature extraction methods with segment-by-segment SVD speech watermarking in the time domain and Wiener filtering

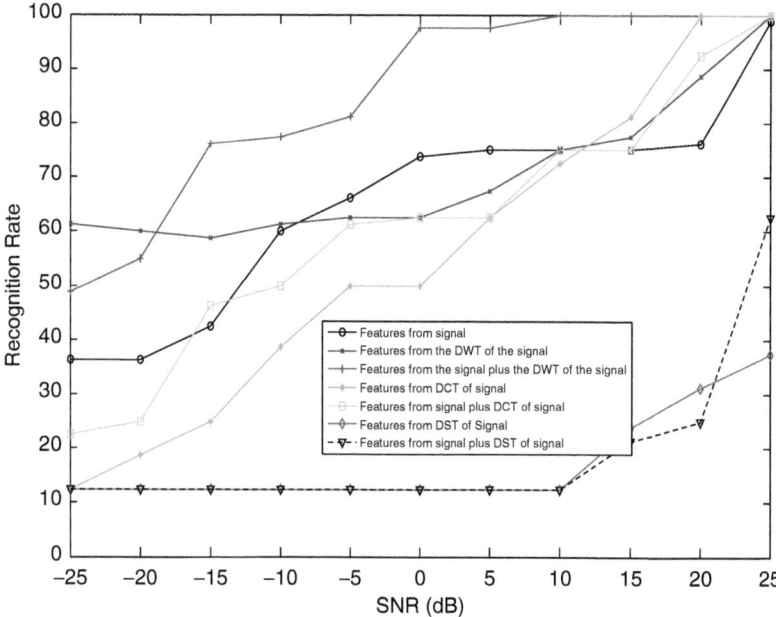

Fig. 86 Recognition rate vs. SNR for the different feature extraction methods with segment-by-segment SVD speech watermarking in the time domain and spectral subtraction

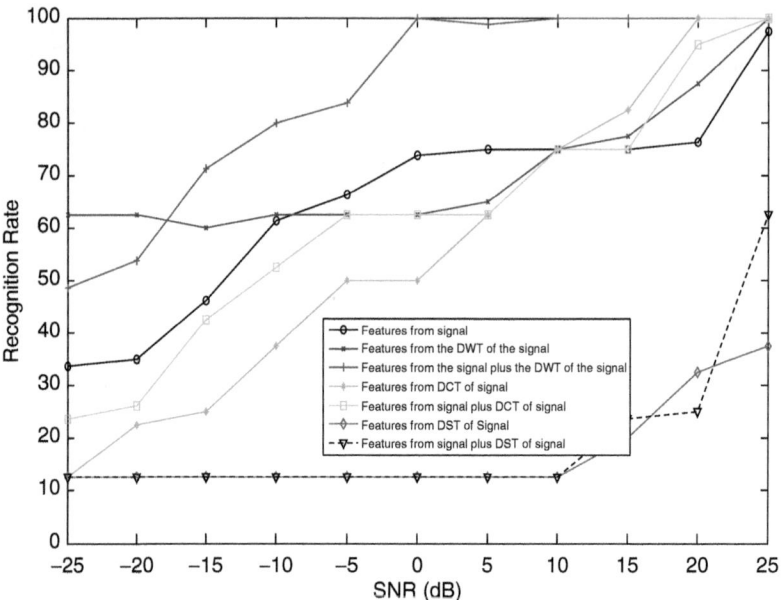

Fig. 87 Recognition rate vs. SNR for the different feature extraction methods with segment-by-segment SVD speech watermarking in the time domain and adaptive Wiener filtering

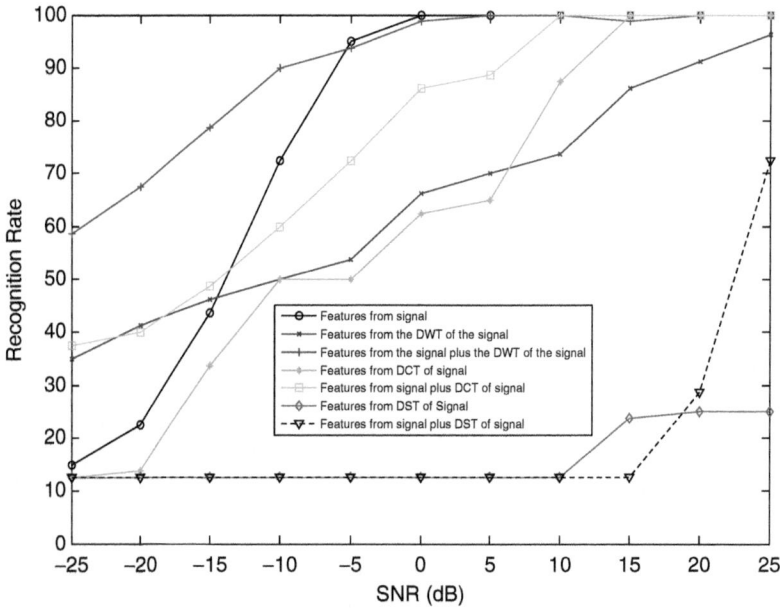

Fig. 88 Recognition rate vs. SNR for the different feature extraction methods with segment-by-segment SVD speech watermarking in the time domain and wavelet soft thresholding with 1 level Haar wavelet

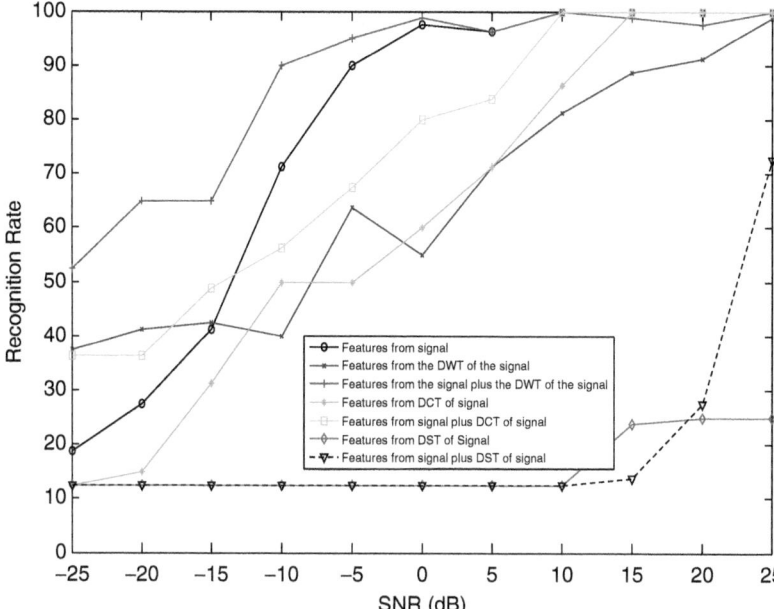

Fig. 89 Recognition rate vs. SNR for the different feature extraction methods with segment-by-segment SVD speech watermarking in the time domain and wavelet hard thresholding with 1 level Haar wavelet

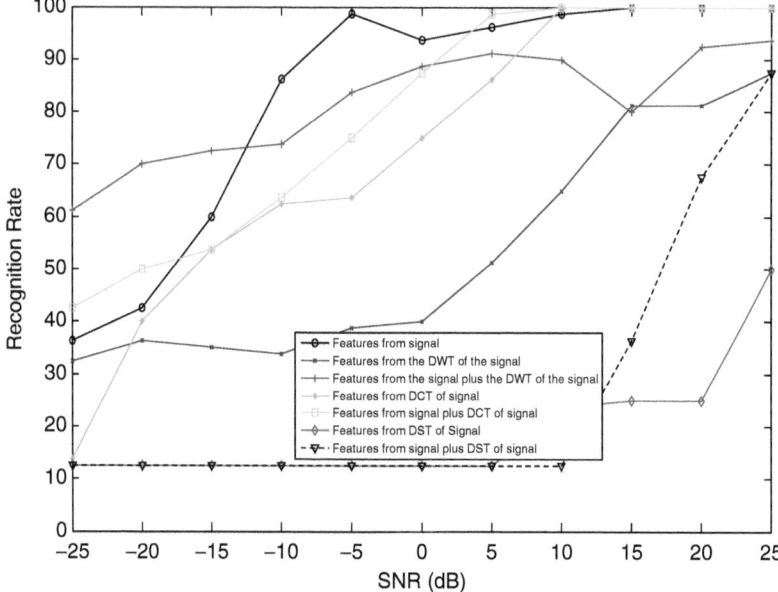

Fig. 90 Recognition rate vs. SNR for the different feature extraction methods with segment-by-segment SVD speech watermarking in the time domain and wavelet soft thresholding with 2 levels Haar wavelet

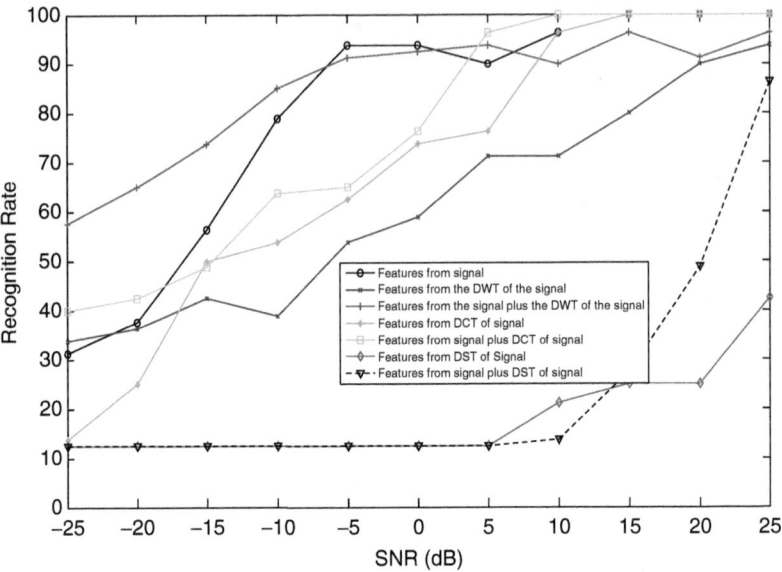

Fig. 91 Recognition rate vs. SNR for the different feature extraction methods with segment-by-segment SVD speech watermarking in the time domain and wavelet hard thresholding with 2 levels Haar wavelet

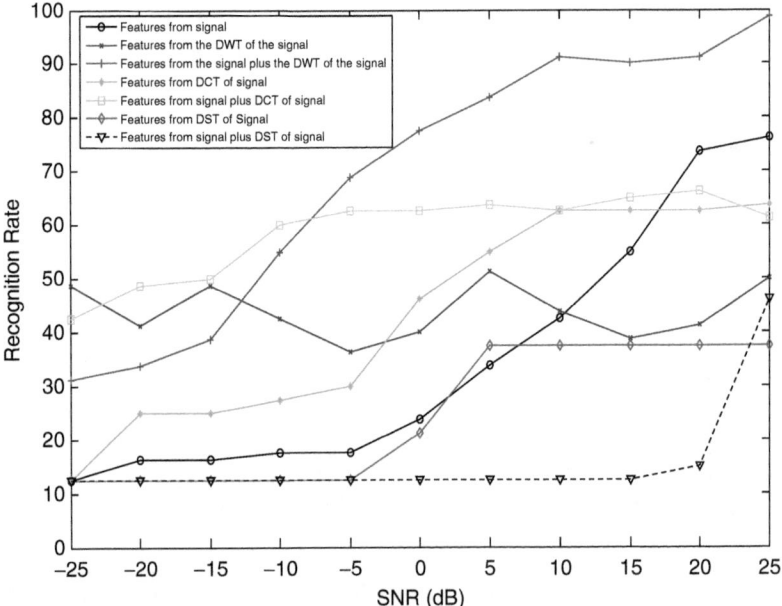

Fig. 92 Recognition rate vs. SNR for the different feature extraction methods with segment-by-segment SVD speech watermarking in the time domain and inverse filter deconvolution

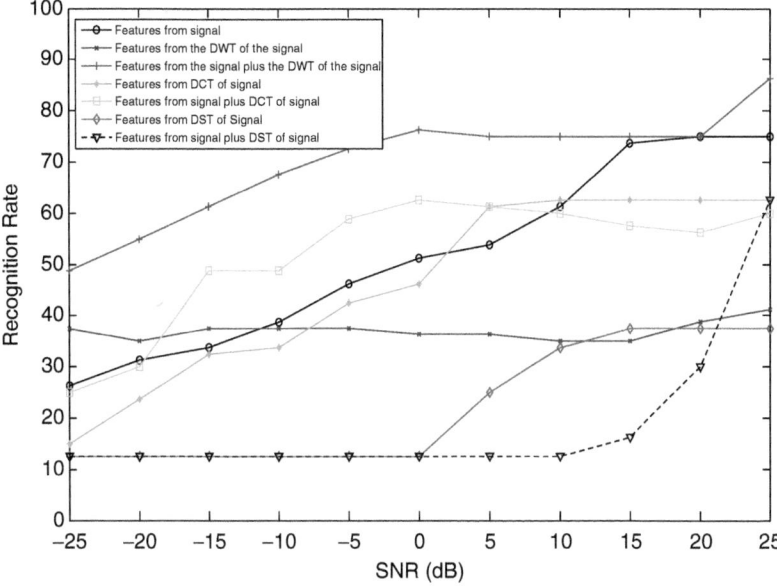

Fig. 93 Recognition rate vs. SNR for the different feature extraction methods with segment-by-segment SVD speech watermarking in the time domain and LMMSE deconvolution

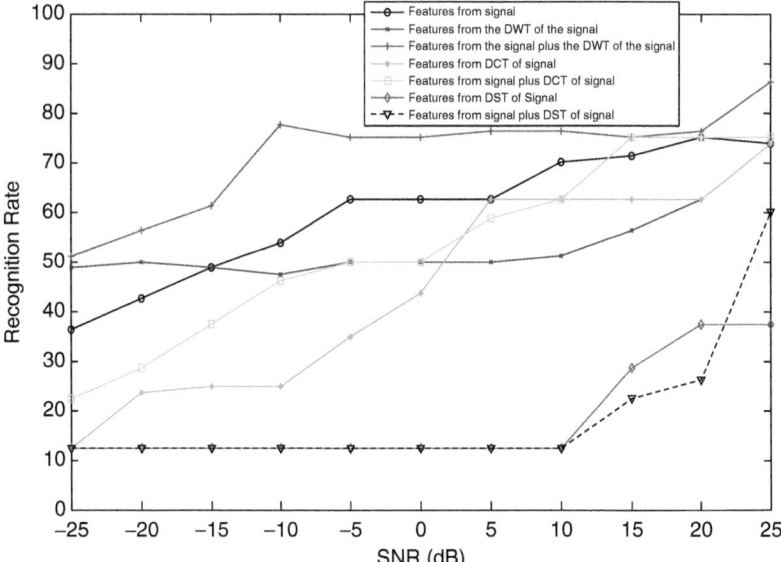

Fig. 94 Recognition rate vs. SNR for the different feature extraction methods with segment-by-segment SVD speech watermarking in the time domain and regularized deconvolution

encryption, the problem is that, if a small portion of the original signal remains intact, it may allow a trained listener to directly interpret the scrambled speech [85].

The objective here is to present an encrypted signal without residual intelligibility in time and frequency domains. To achieve this objective, speech encryption can be performed with multiple secret keys that are used for the permutation and masking of speech segments in both time and transform domains. The encryption steps can be summarized as follows [86]:

1. Framing and reshaping into 2D blocks.
2. Block randomization.
3. First round:

   - Generation of key 1.
   - Permutation with key 1.
   - Generation of mask 1.
   - Addition of mask 1.

4. Second round:

   - DCT or DST.
   - Generation of key 2.
   - Permutation with key 2.
   - Generation of mask 2.
   - Addition of mask 2.

5. Third round:

   - IDCT, or inverse DST (IDST).
   - Generation of key 3.
   - Permutation with key 3.

6. Reshaping into 1D format.

The decryption steps can be summarized as follows [86]:

- Generation of key 1.
- Generation of key 2.
- Generation of key 3.
- Framing and reshaping into 2D blocks.
- Inverse permutation with key 3.
- DCT or DST.
- Generation of mask 2.
- Subtraction of mask 2.
- Inverse permutation with key 2.
- IDCT or IDST.
- Generation of mask 1.
- Subtraction of mask 1.
- Inverse permutation with key 1.
- Inverse of the block randomization process.

8.1   *Framing and Reshaping into 2D Blocks*

Continuous speech signals are sampled and recorded in sound files in the form of streams of discrete speech samples with amplitudes between -1 and 1. The series of samples are framed and reshaped into square blocks with width equal to the secret key length.

8.2   *Block Randomization*

Block randomization is performed with circular shifts as shown in Fig. 95. The first row remains intact. The second row is circularly shifted single step to the right. The third row is circularly shifted two steps to the right. Similar shifts are performed for the other rows.

a

A_1	A_6	A_{11}	A_{16}	A_{21}
A_2	A_7	A_{12}	A_{17}	A_{22}
A_3	A_8	A_{13}	A_{18}	A_{23}
A_4	A_9	A_{14}	A_{19}	A_{24}
A_5	A_{10}	A_{15}	A_{20}	A_{25}

Original Block.

b

No action		A_1	A_6	A_{11}	A_{16}	A_{21}
One steps cyclic shift		A_{22}	A_2	A_7	A_{12}	A_{17}
Two steps cyclic shift		A_{18}	A_{23}	A_3	A_8	A_{13}
Three steps cyclic shift		A_{14}	A_{19}	A_{24}	A_4	A_9
Four steps cyclic shift		A_{10}	A_{15}	A_{20}	A_{25}	A_5

Randomized Block.

Fig. 95   Block randomization. (**a**) Original block. (**b**) Randomized block

8.3 Steps of the Rounds

8.3.1 Generation of Keys

The first step is the generation of the original secret key. It can be generated by a PN sequence generator. This secret key is shared between the transmitter and receiver. The second key is the inverse of the original key. The third key is generated from the original key by dividing it into two halves and reversing the two halves. An example of three keys is key 1 = 11001000, key 2 = 00110111, and key 3 = 10001100.

8.3.2 Permutation with a Key

The generated keys control the permutation process. The first key is applied to the rows of the resulting randomized block. If a key bit equals 1, the whole corresponding row is circularly shifted to the right by a number of shifts equal to the row number minus one (e.g., row number 14 is shifted 13 times). If a key bit equals 0, the corresponding row remains intact as shown in Fig. 96. After that, the same key is applied to the columns of the resulting block in a similar manner as shown in Fig. 97. The two other keys are used in a similar manner in the subsequent rounds.

8.4 Masking

Permutation of speech segments in time domain results in a distortion of the speech time envelope, which reduces the intelligibility of the speech. However, some

Key bits		Block before permutation						Block after permutation				
1	No action	B_1	B_6	B_{11}	B_{16}	B_{21}		B_1	B_6	B_{11}	B_{16}	B_{21}
1	one step cyclic shift	B_2	B_7	B_{12}	B_{17}	B_{22}	➡	B_{22}	B_2	B_7	B_{12}	B_{17}
0	No action	B_3	B_8	B_{13}	B_{18}	B_{23}		B_3	B_8	B_{13}	B_{18}	B_{23}
1	Three steps cyclic shift	B_4	B_9	B_{14}	B_{19}	B_{24}		$B1_4$	B_{19}	B_{24}	B_4	B_9
1	Four steps cyclic shift	B_5	B_{10}	B_{15}	B_{20}	B_{25}		B_{10}	B_{15}	B_{20}	B_{25}	B_5

Fig. 96 Row permutation step

Key bits										
1	**1**	**0**	**1**	**1**						
No action	One step cyclic shift	No action	Three steps cyclic shift	Four steps cyclic shift						
C_1	C_6	C_{11}	C_{16}	C_{21}		C_1	C_{10}	C_{11}	C_{18}	C_{22}
C_2	C_7	C_{12}	C_{17}	C_{22}		C_2	C_6	C_{12}	C_{19}	C_{23}
C_3	C_8	C_{13}	C_{18}	C_{23}		C_3	C_7	C_{13}	C_{20}	C_{24}
C_4	C_9	C_{14}	C_{19}	C_{24}		C_4	C_8	C_{14}	C_{16}	C_{25}
C_5	C_{10}	C_{15}	C_{20}	C_{25}		C_5	C_9	C_{15}	C_{17}	C_{21}
Block before column permutation.						Block after column permutation.				

Fig. 97 Column permutation step

portions of the signal remain intact, which may allow a trained listener to directly interpret the scrambled speech. Therefore, a masking step is very necessary in order to change the remaining nonpermutated portions of speech signals and to increase the security of the cryptosystem.

The utilized mask is generated from the key using a number of circular shifts of the key equal to the number of sample rows minus one as shown in Fig. 98. The resultant mask is added to each block of samples as shown in Fig. 98. After the mask addition, a value of 2 is subtracted from all values exceeding 1 resulting in negative values. In the decryption process, the mask is subtracted from each block, and a value of two is added to all values below -1 to guarantee the correct reconstruction of the sample values.

8.5   Discrete Transforms

The objective of using either the DCT or DST is to remove the residual intelligibility of speech signals after the masking step. Each of these transforms has a strong diffusion mechanism. All samples in time domain contribute to each sample in the transform domain, which guarantees a totally different shape of the transformed signals. Another permutation step is performed on the transform domain samples to increase the security prior to the inversion of the transform and the application of another permutation step in the time domain.

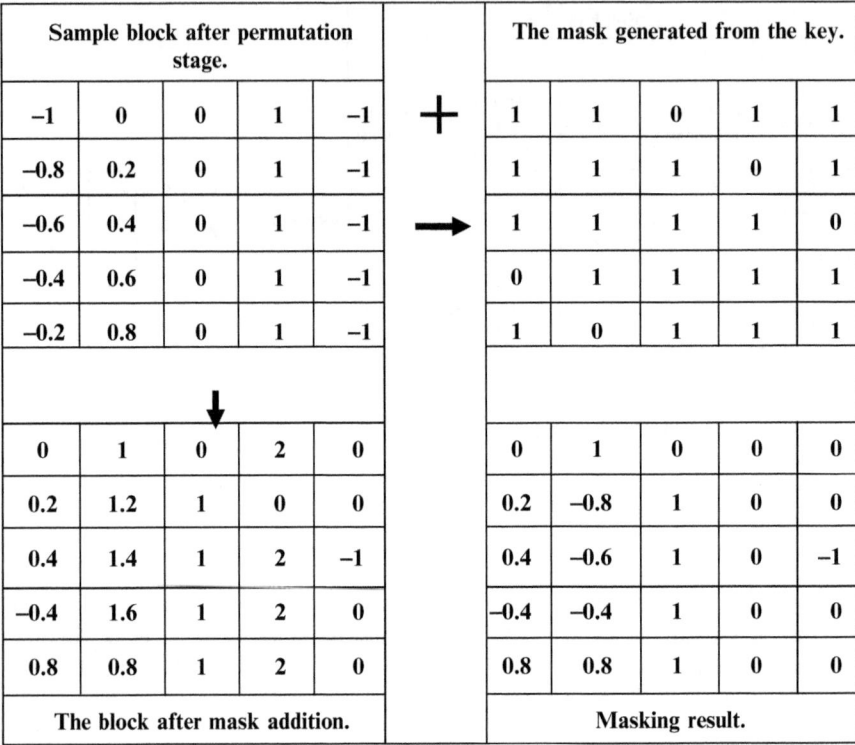

Fig. 98 Masking step

8.6   Performance Evaluation of Speech Encryption

Several experiments have been carried out to test the encryption efficiency of the speech cryptosystem. The qualities of both the encrypted and decrypted speech signals have been assessed. The speech signal used in all experiments is shown in Fig. 99a. It is a synthetic signal for the sentence "We were away year ago." The first 2.5 s are for a female saying this sentence. The next 1.5 s are a perfect silence period without noise. The next 1.5 s are for a silence period with room noise. The last 2.5 s are for a male saying the same sentence. This signal is encrypted with the proposed cryptosystem in the time domain using a single round only, and the result is shown in Fig. 99b. TD refers to time domain. The spectrograms of the original and encrypted signals are shown in Fig. 100. It is evident that the encrypted speech with the DCT and DST encryption is obviously similar to the white noise without any talk spurts. The original intonations have been removed, which indicates that no residual intelligibility can be useful for eavesdroppers at the communication channel.

The different kinds of ciphers can be analyzed, statistically [87–89]. Statistical analysis has been performed on the above-mentioned cryptosystem demonstrating

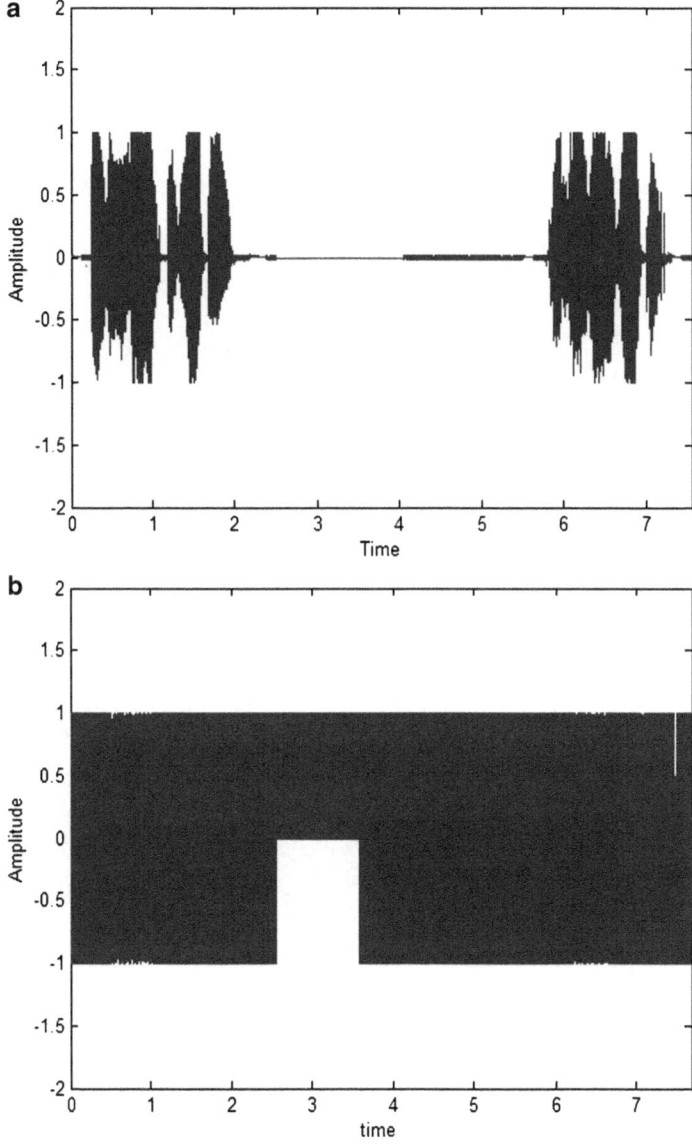

Fig. 99 Encryption of the speech signal. (**a**) Original signal. (**b**) TD encryption. (**c**) DCT encryption. (**d**) DST encryption

its superior confusion and diffusion properties, which strongly resist the statistical attacks. This is illustrated by showing the correlation coefficient between encrypted signal and the original signal and the SD of the encrypted signal compared to the original one. The correlation coefficients between the encrypted speech signals and the original speech signal for all methods using three different main keys are

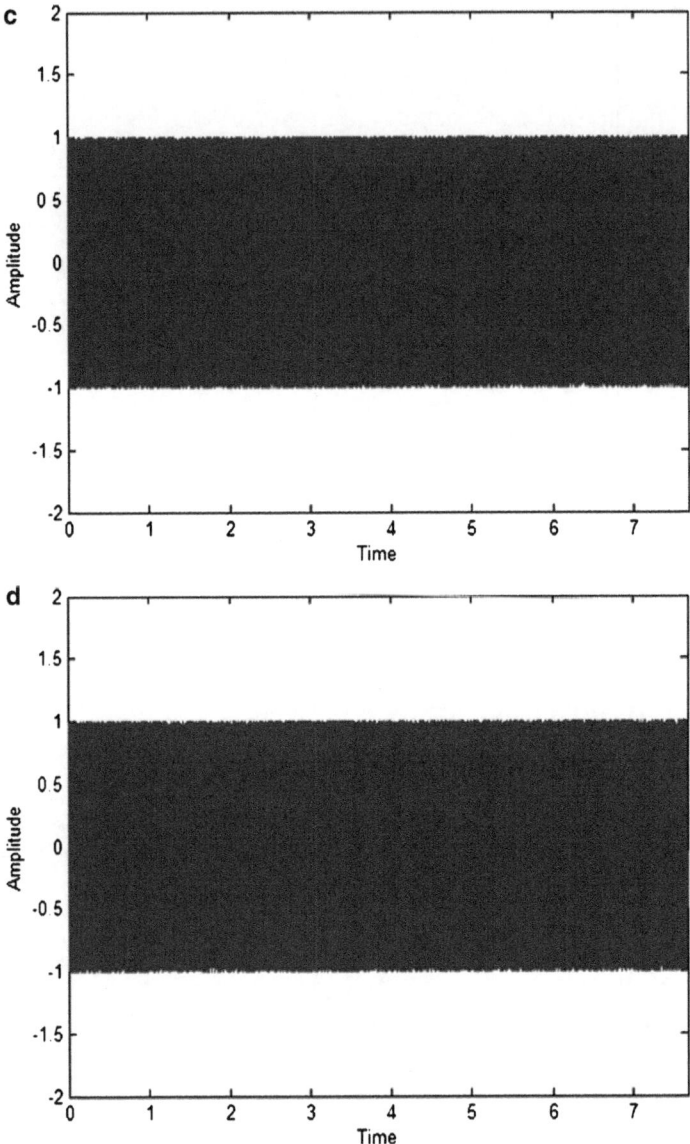

Fig. 99 (continued)

tabulated in Table 3. From these results, we can see that all secret keys produce encrypted speech signals with low correlation between similar segments in the original speech and the encrypted speech, which means that all keys give good encryption results. The SD results for the encrypted signals for all methods are tabulated in Table 4.

A secure encryption algorithm should be sensitive to the cipher keys. For the speech cryptosystem, the key space analysis and sensitivity test have been performed. For a secure cryptosystem, the key space should be large enough to make the brute-force attack infeasible [89]. An exhaustive key search needs 2^{s_k} operations to succeed, where s_k is the key size in bits. An attacker simply tries all

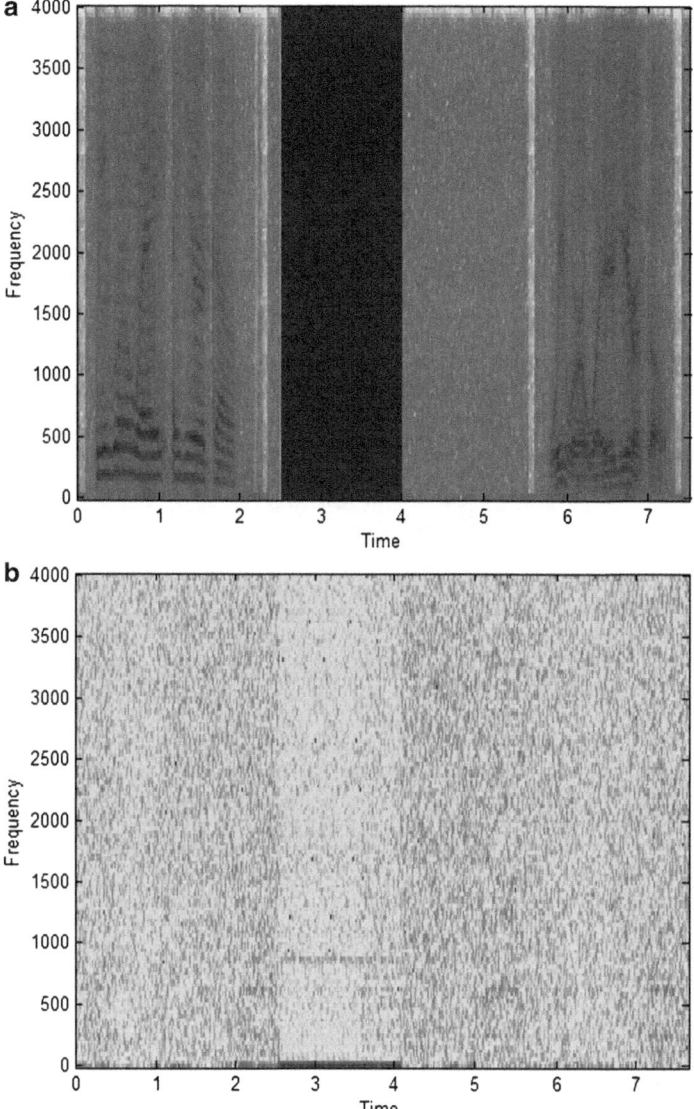

Fig. 100 Spectrograms of speech signals (**a**) Original signal. (**b**) TD encryption. (**c**) DCT encryption. (**d**) DST encryption

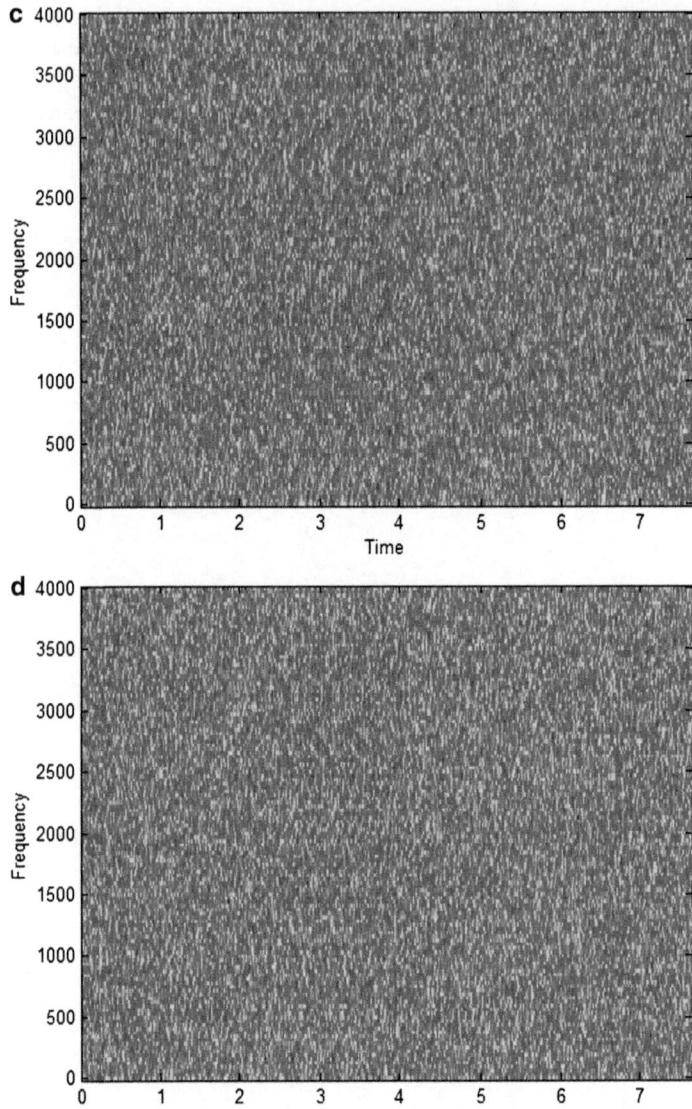

Fig. 100 (continued)

Table 3 Correlation coefficients between the original and encrypted speech signals

Secret key	TD	DCT	DST
Key A	0.0035	0.0043	0.0050
Key B	0.0048	0.0012	−0.0015
Key C	0.0039	0.0014	0.0036

Table 4 SD in dB of the encrypted signals with all methods

Secret key	TD	DCT	DST
Key A	26.1445	23.1539	22.9874
Key B	27.1769	22.1501	22.1960
Key C	26.7445	23.1238	21.9133

Table 5 Correlation coefficients between the decrypted signals with the different keys and the decrypted signal with the original key

Decryption key	TD	DCT	DST
Key 1	0.1757	0.0875	0.0208
Key 2	0.0644	0.0220	0.0923
Key 3	0.0130	0.0081	0.0202

keys and this is very exhaustive. Assuming that, the secret key length is 128 bits, therefore; an opponent needs about 2^{128} operations to successfully determine the key. If the opponent employs 1,000 million instructions per second (MIPS) to guess the key, the computations require:

$$\frac{2^{128}}{365 \times 24 \times 60 \times 60 \times 1,000 \times 10^6} > 10.7902831 \times 10^{21} \text{ years.}$$

For a 64-bits key, the computations require:

$$\frac{2^{64}}{365 \times 24 \times 60 \times 60 \times 1,000 \times 10^6} > 584 \text{ years.}$$

These results suppose a known secret key length by the attacker, but really the key length is unknown making the search infeasible.

Key sensitivity means that the encrypted signal cannot be decrypted correctly, if there is any change between encryption and decryption keys [81]. Large key sensitivity is required by all secure cryptosystems. Assume that a key consisting of 64 bits is used for encryption. For testing the key sensitivity of the proposed cryptosystem, the encrypted signal is decrypted with three different keys generated by changing only a single bit in the original secret key. The correlation coefficient is estimated between each decrypted signal and the signal decrypted with the original key, and the results are tabulated in Table 5. The low correlation values show the large key sensitivity of the speech cryptosystem implementing the DCT or the DST.

The known-plaintext attack is an attack model for cryptanalysis, where the attacker has samples of both the plaintext and its ciphertext and has liberty to make use of them to reveal the secret key. In the speech cryptosystem, if a cryptanalyst knows the original signal and its encrypted version, he must know the block size to build the permutation and masking processes. If he tries with a different block size, this will give completely wrong results. In modern cryptosystems that use standard block sizes, permutation and substitution processes may be analyzed to discover the key, while in the above-mentioned speech cryptosystem; there is no standard block size. Therefore, the knowledge of the plaintext without knowledge of the block size is useless, as it is very difficult to guess the key.

Three metrics have been used for quality assessment of decrypted speech signals; the SD, the LLR, and the correlation coefficient with the original speech signal. As the values of the SD and the LLR decrease, and the value of the correlation coefficient increases, the performance of the speech cryptosystem becomes better. Figure 101 shows the decrypted signals with all methods in the absence of noise. The numerical quality metrics values for these results are tabulated in Table 6. These results ensure the efficiency of the speech cryptosystem in the absence of noise.

An important issue, which deserves consideration, is the effect of noise on the efficiency of the speech cryptosystem. Simulation experiments have been carried out for the decryption in the presence of noise at different SNR values. The results of these experiments are shown in Figs. 102–104 for all encryption methods. From these results, it is clear that the encryption quality metrics values are better at high SNR values. Thus, the speech cryptosystem can tolerate noise with high SNR values.

Encryption of speech signals prior to transmission in remote access speaker identification system is recommended to increase the degree of security. The effect of encryption on the performance of the speaker identification system has been studied and the results are given in Figs. 105–113. These results reveal that speech encryption can increase the system security without any degradation in the speaker identification system performance.

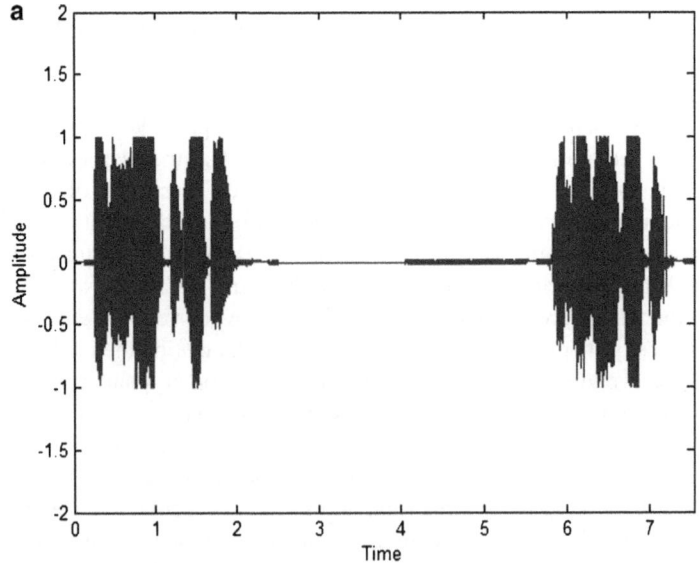

Fig. 101 Decrypted speech signals (**a**) TD. (**b**) DCT. (**c**) DST

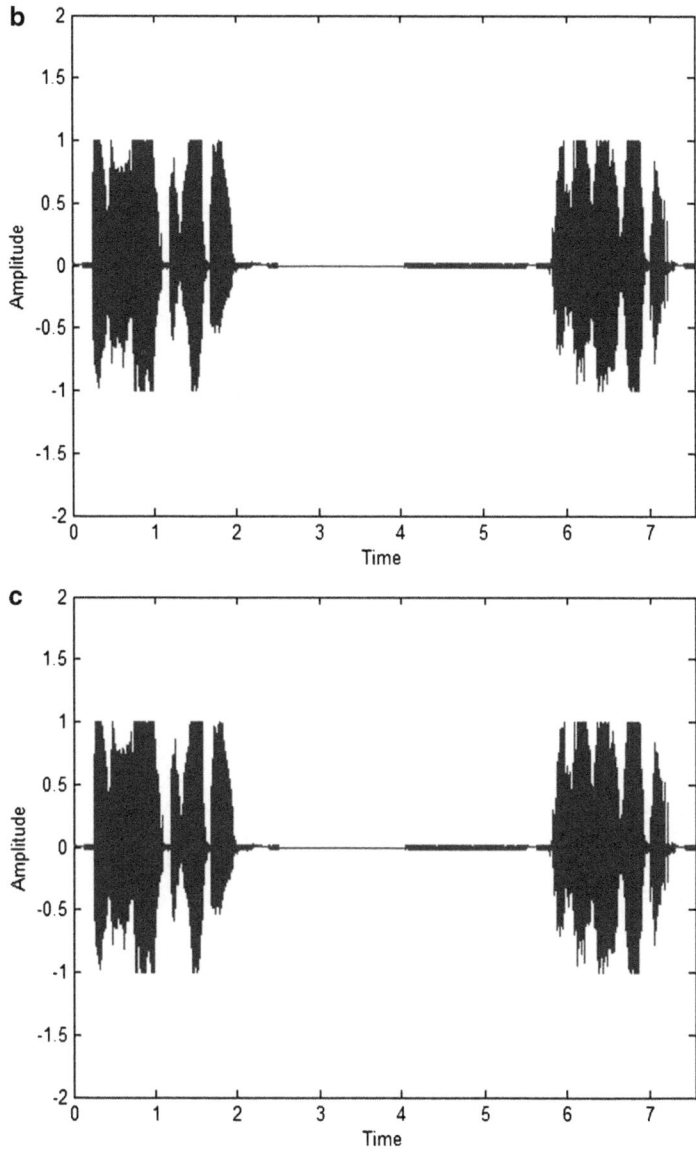

Fig. 101 (continued)

Table 6 Quality metrics values for the decrypted speech signals

Quality metrics	TD	DCT	DST
SD	0.044	0.044	0.044
LLR	8.89E−8	8.89E−8	8.89E−8
r_{xz}	0.9998	0.9997	0.9993

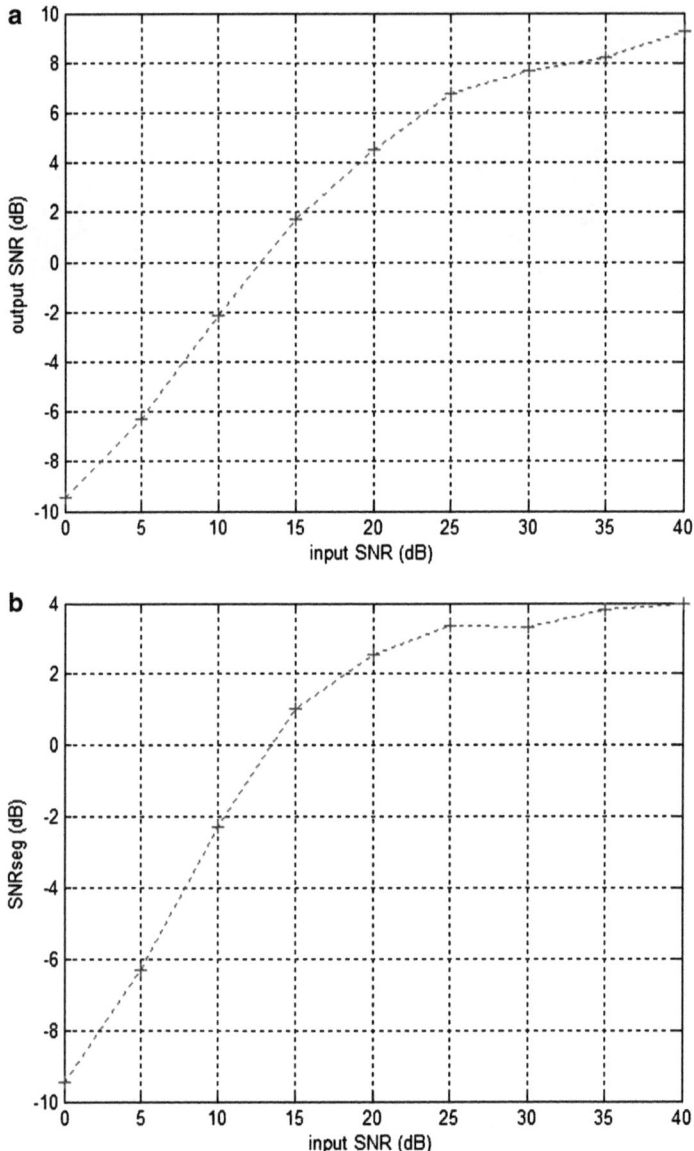

Fig. 102 Variation of the quality metrics values of the decrypted signal with the SNR in the presence of noise for TD encryption. (**a**) SNR. (**b**) SNRseg. (**c**) LLR. (**d**) SD

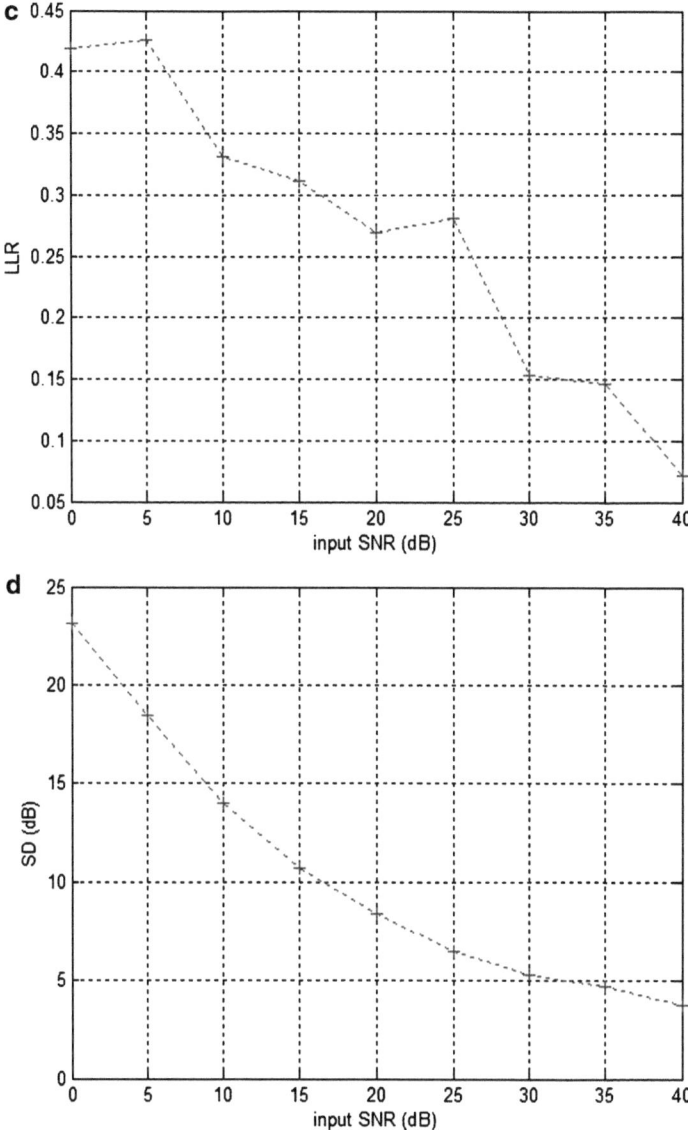

Fig. 102 (continued)

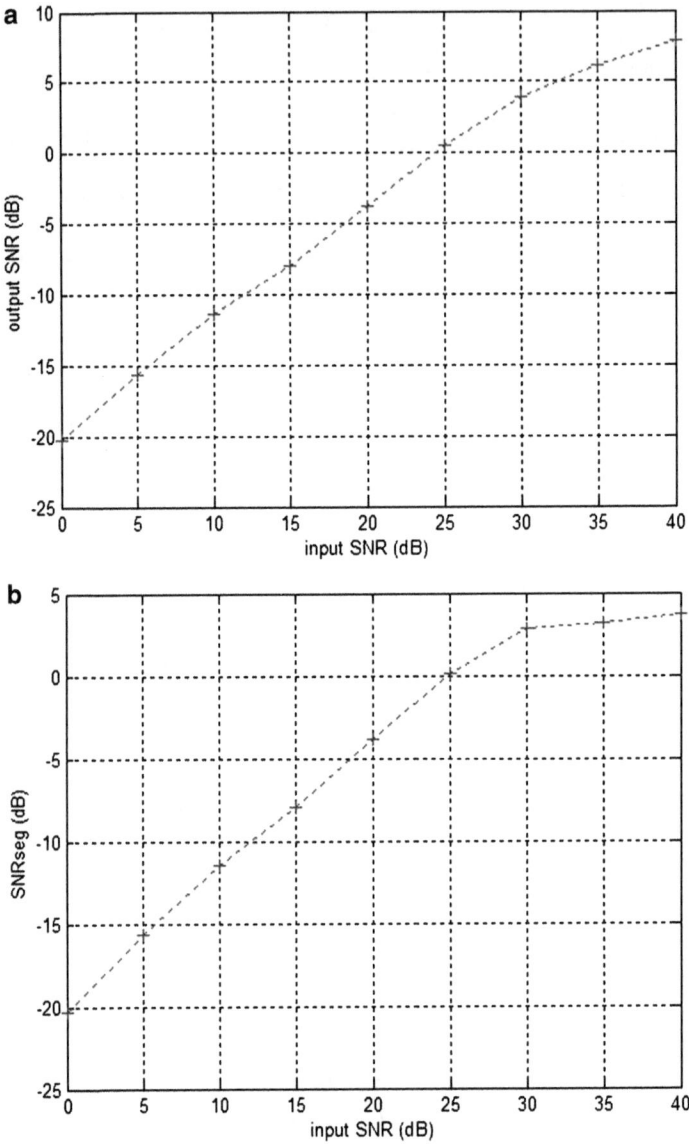

Fig. 103 Variation of the quality metrics values of the decrypted signal with the SNR in the presence of noise for DCT encryption. (**a**) SNR. (**b**) SNRseg. (**c**) LLR. (**d**) SD

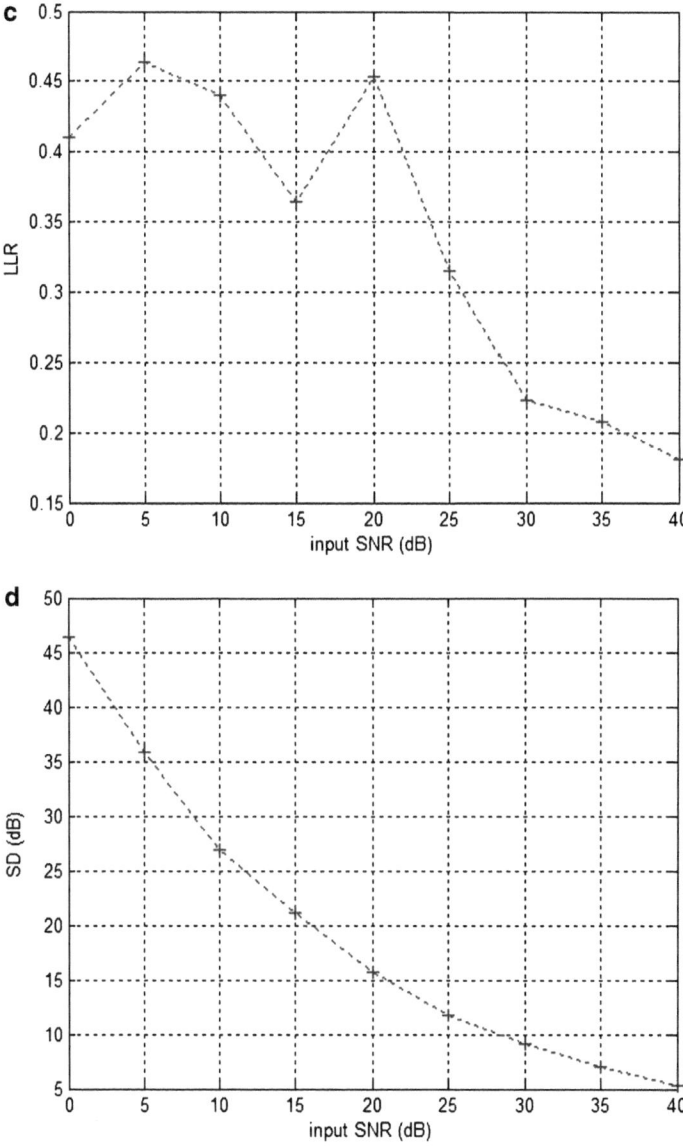

Fig. 103 (continued)

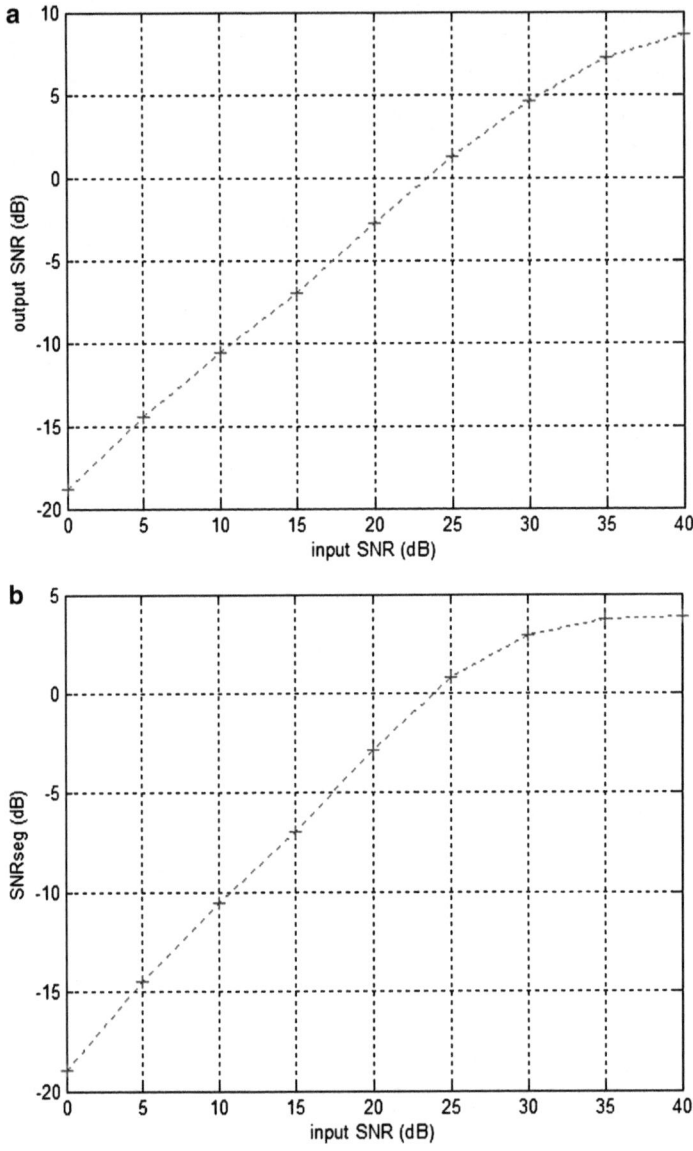

Fig. 104 Variation of the quality metrics values of the decrypted signal with the SNR in the presence of noise for DST encryption. (**a**) SNR. (**b**) SNRseg. (**c**) LLR. (**d**) SD

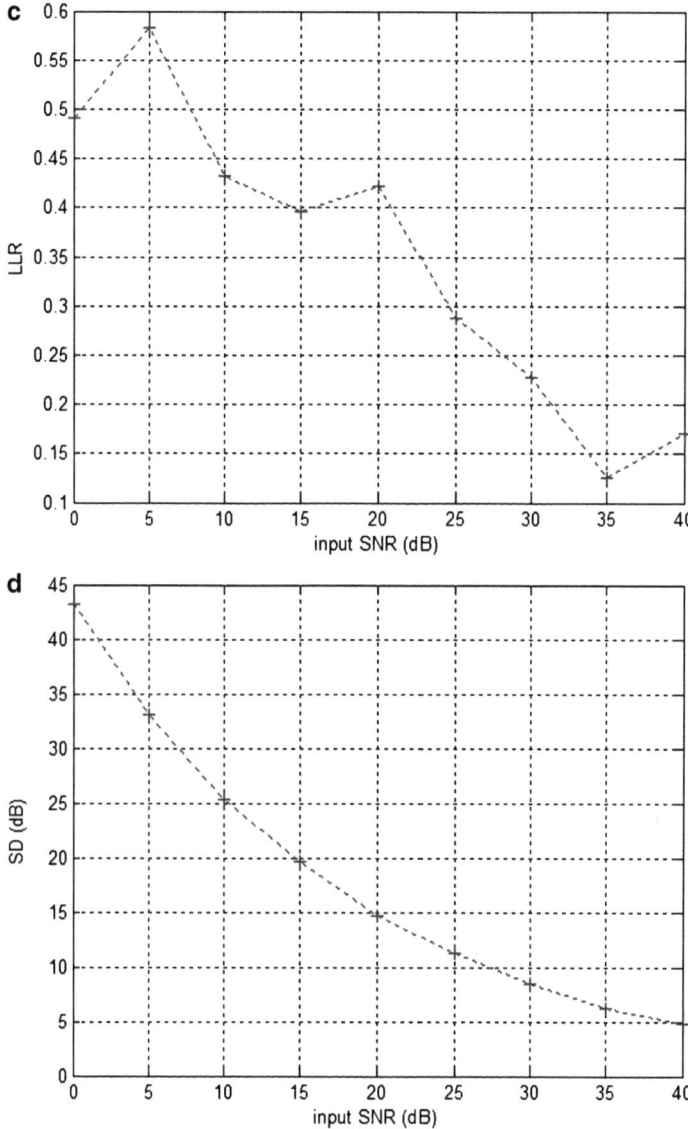

Fig. 104 (continued)

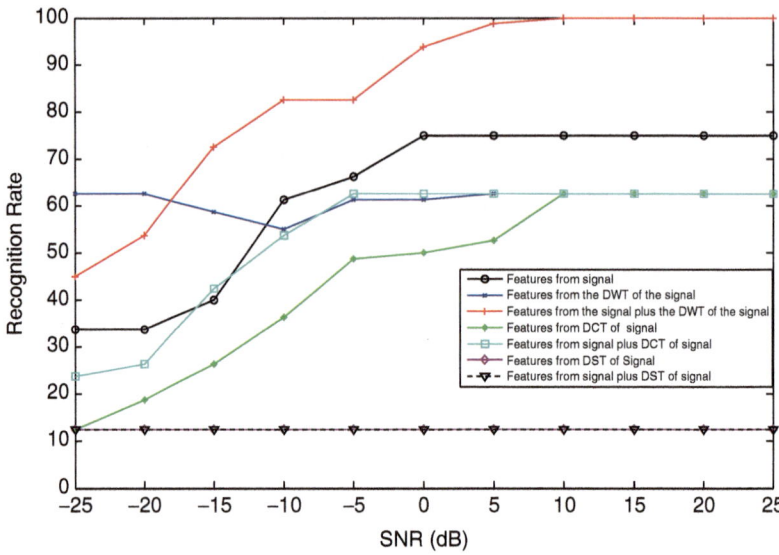

Fig. 105 Recognition rate vs. SNR for the different feature extraction methods with speech encryption and AWGN

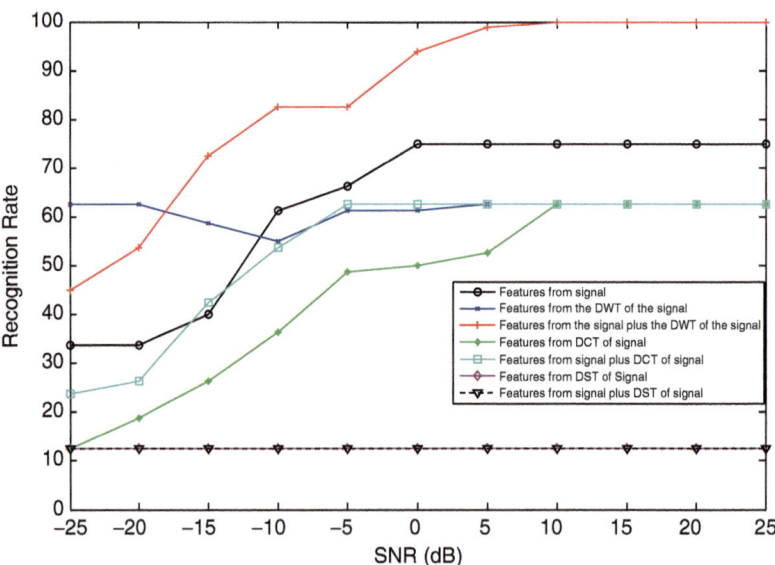

Fig. 106 Recognition rate vs. SNR for the different feature extraction methods with speech encryption and Wiener filtering

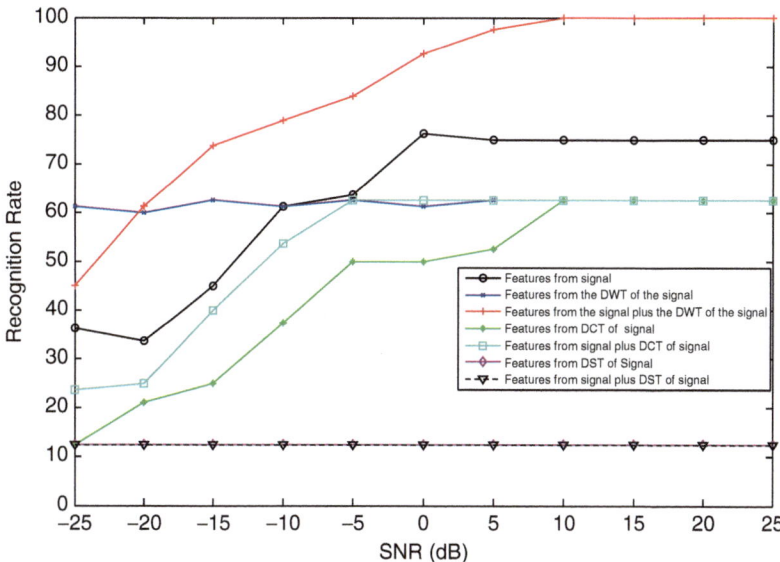

Fig. 107 Recognition rate vs. SNR for the different feature extraction methods with speech encryption and spectral subtraction

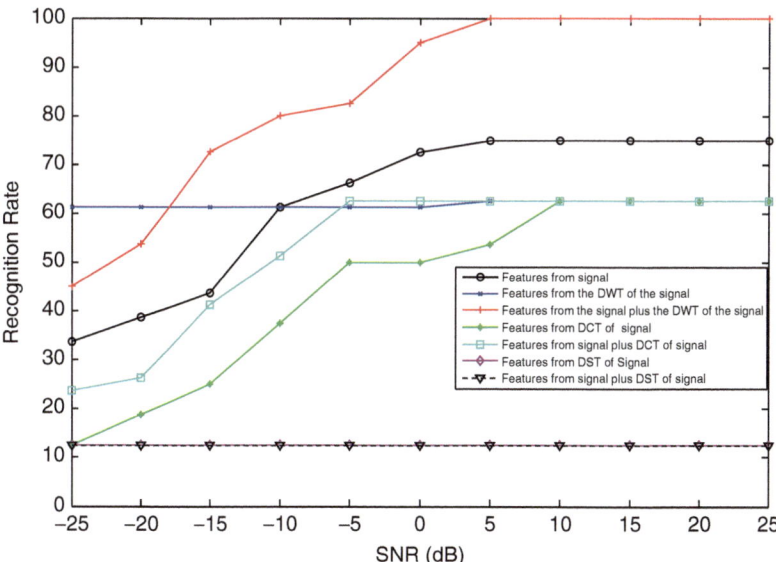

Fig. 108 Recognition rate vs. SNR for the different feature extraction methods with speech encryption and adaptive Wiener filtering

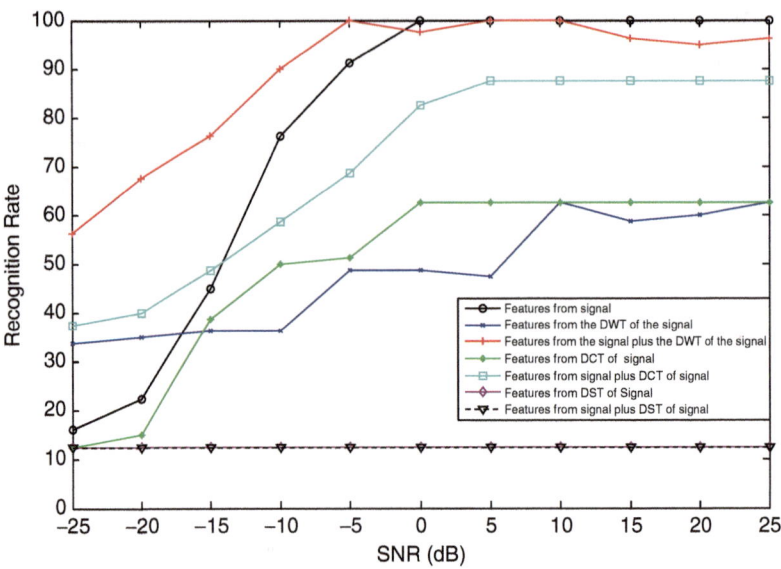

Fig. 109 Recognition rate vs. SNR for the different feature extraction methods with speech encryption and wavelet denoising with 1 level Haar wavelet

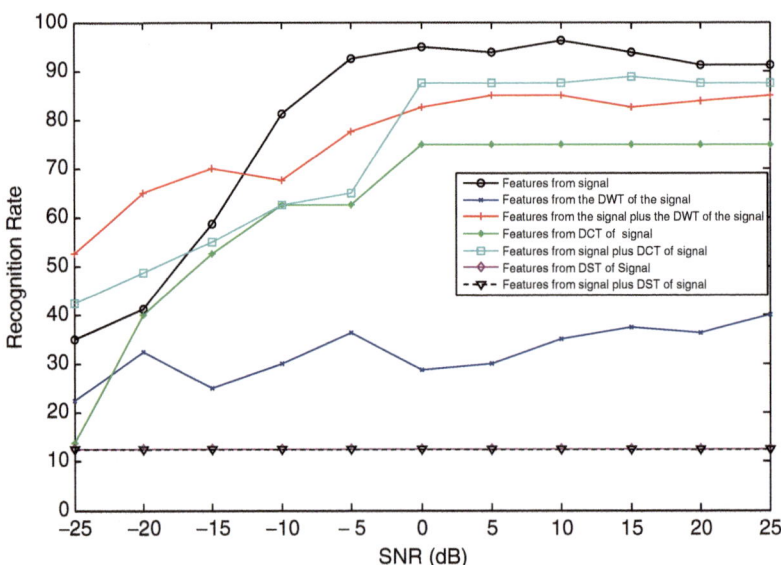

Fig. 110 Recognition rate vs. SNR for the different feature extraction methods with speech encryption and wavelet denoising with 2 levels Haar wavelet

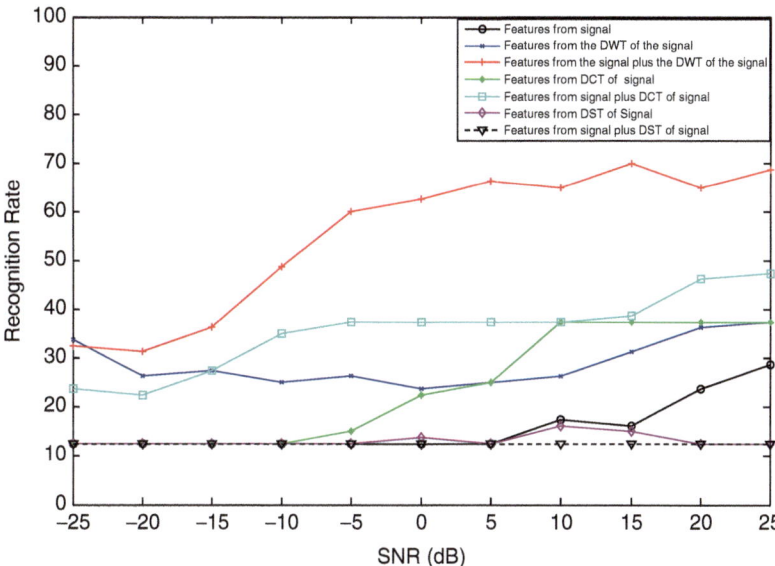

Fig. 111 Recognition rate vs. SNR for the different feature extraction methods with speech encryption and inverse filter deconvolution

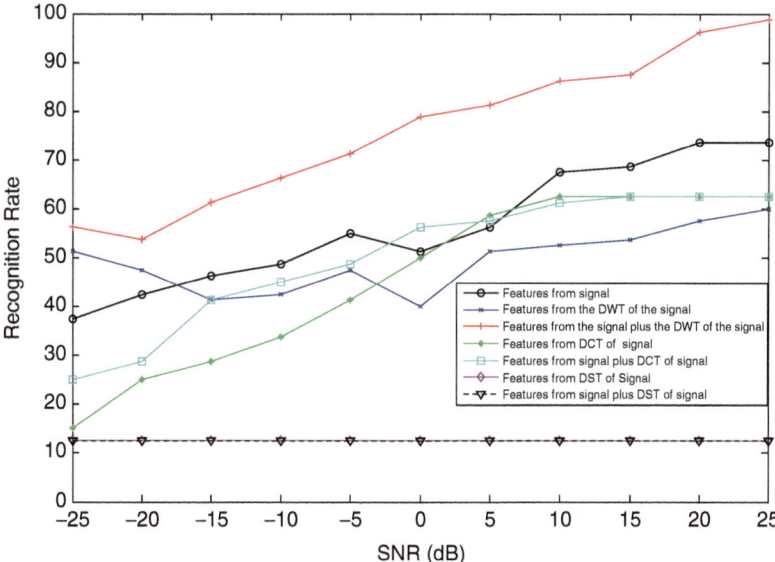

Fig. 112 Recognition rate vs. SNR for the different feature extraction methods with speech encryption and LMMSE deconvolution

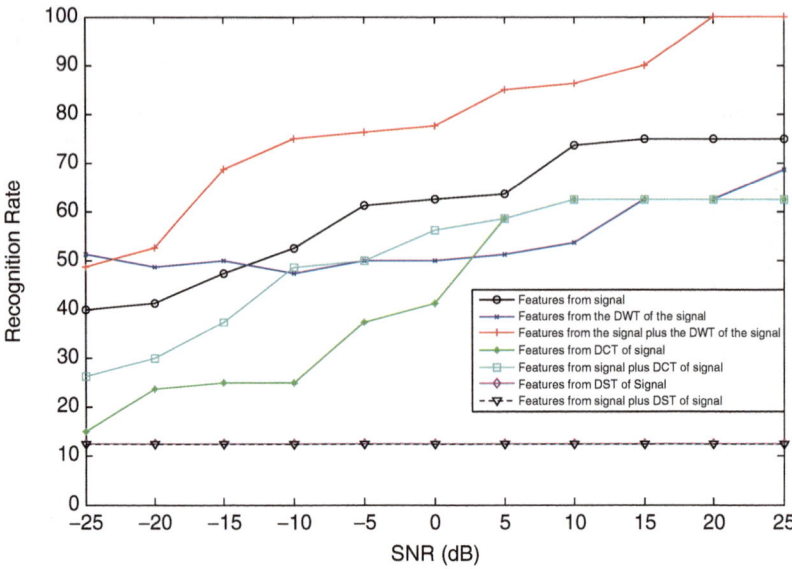

Fig. 113 Recognition rate vs. SNR for the different feature extraction methods with speech encryption and regularized deconvolution

9   Simultaneous Watermarking and Encryption

Both speech watermarking and encryption of speech signals can be incorporated with the speaker identification system prior to transmission in remote access speaker identification systems to achieve three levels of security. The effect of watermarking and encryption on the performance of the speaker identification system has been studied and the results are given in Figs. 114–122. These results reveal that speech watermarking and encryption can be used in high-level authentication systems, without any noticeable degradation in the speaker identification process.

10   Conclusion

This book presented a literature survey on speaker identification systems that are based on MFCCs and neural matching. It presented a study for the performance of these systems in noisy environments. The different transform domains were investigated in the book for robust feature extraction in the presence of noise. Simulation results revealed that feature extraction from the DCT or the DWT of speech signals is very feasible for performance enhancement of speaker identification systems. The book developed a new implementation of speech enhancement,

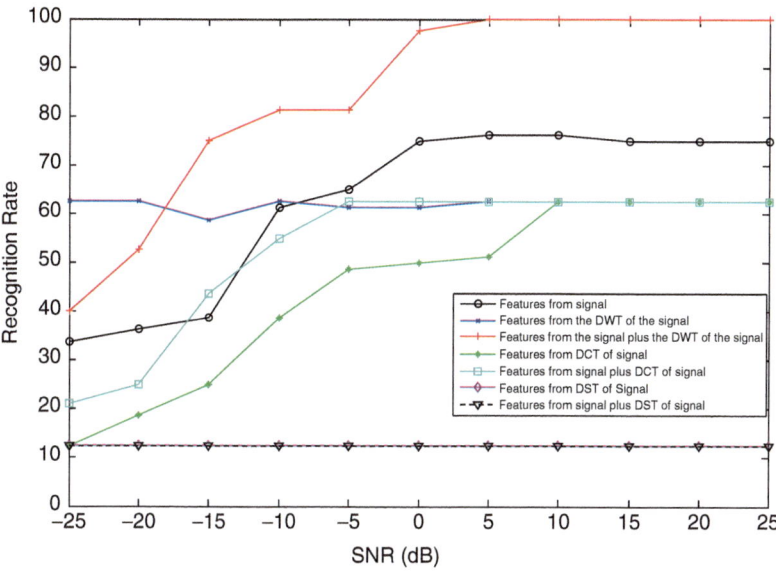

Fig. 114 Recognition rate vs. SNR for the different feature extraction methods with segment-by-segment SVD speech watermarking, speech encryption, and AWGN

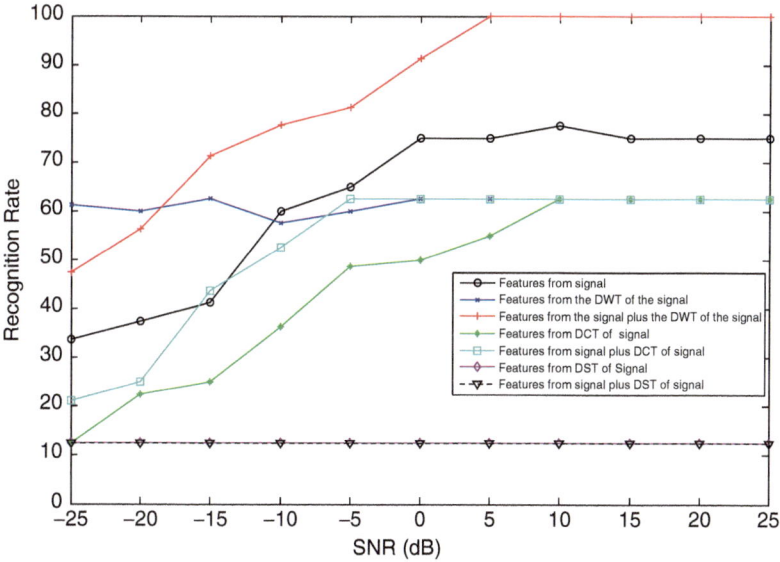

Fig. 115 Recognition rate vs. SNR for the different feature extraction methods with segment-by-segment SVD speech watermarking, speech encryption, and Wiener filtering

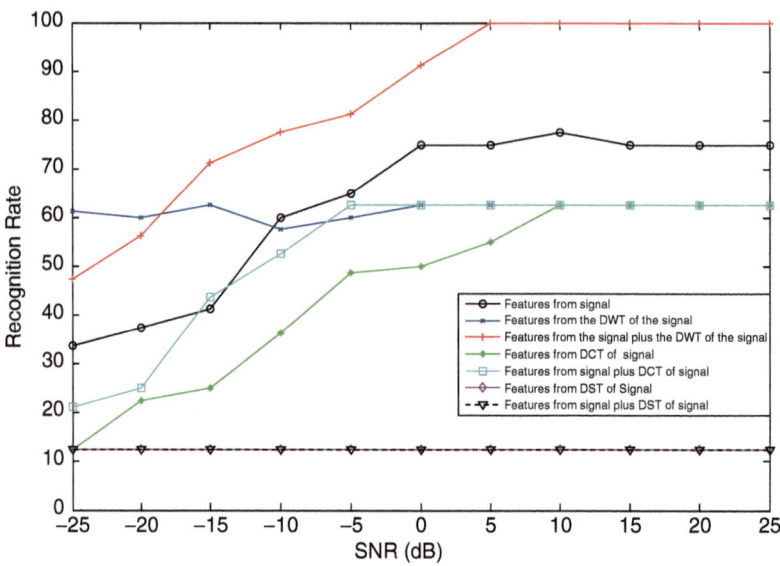

Fig. 116 Recognition rate vs. SNR for the different feature extraction methods with segment-by-segment SVD speech watermarking, speech encryption, and spectral subtraction

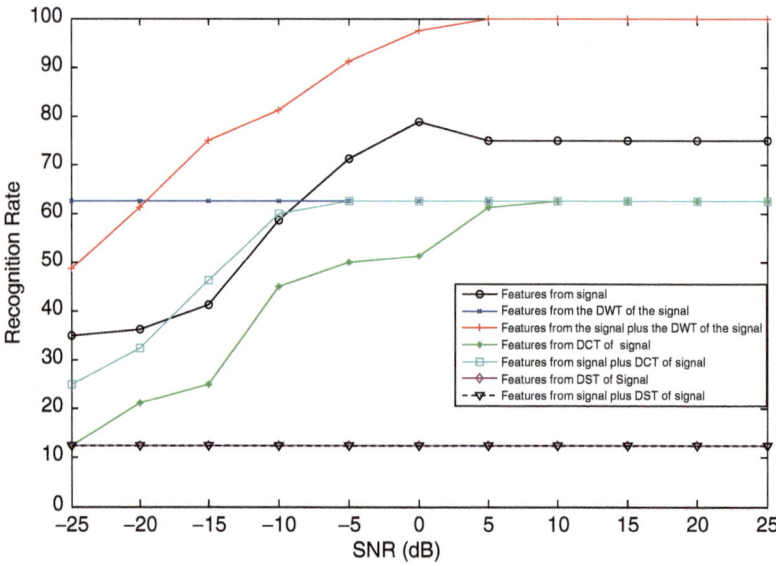

Fig. 117 Recognition rate vs. SNR for the different feature extraction methods with segment-by-segment SVD speech watermarking, speech encryption, and adaptive Wiener filtering

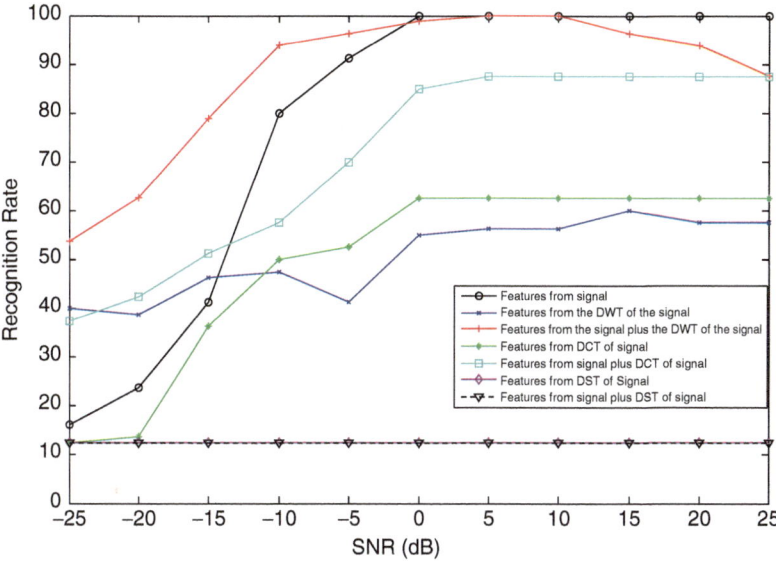

Fig. 118 Recognition rate vs. SNR for the different feature extraction methods with segment-by-segment SVD speech watermarking, speech encryption, and wavelet denoising with 1 level Haar wavelet

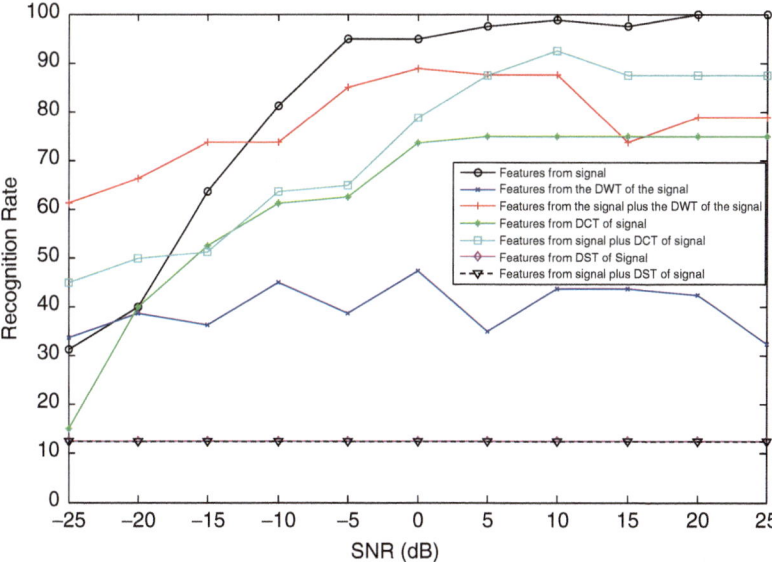

Fig. 119 Recognition rate vs. SNR for the different feature extraction methods with segment-by-segment SVD speech watermarking, speech encryption, and wavelet denoising with 2 levels Haar wavelet

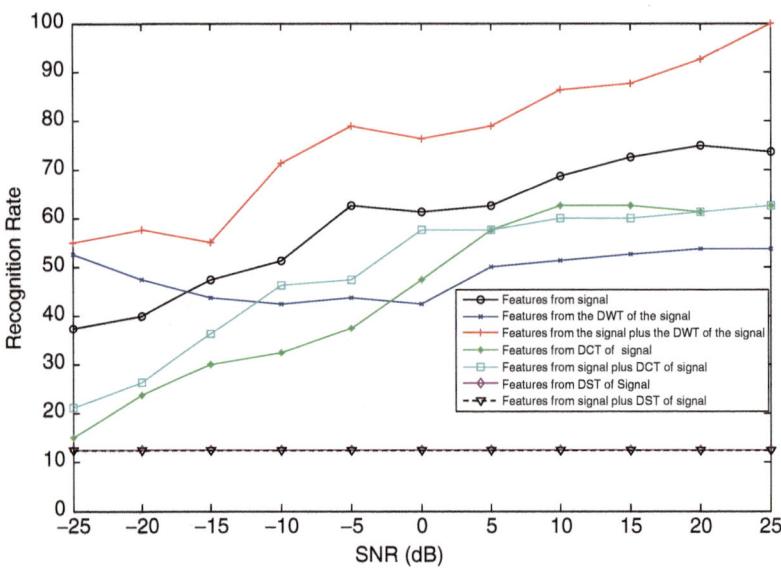

Fig. 120 Recognition rate vs. SNR for the different feature extraction methods with segment-by-segment SVD speech watermarking, speech encryption, and inverse filter deconvolution

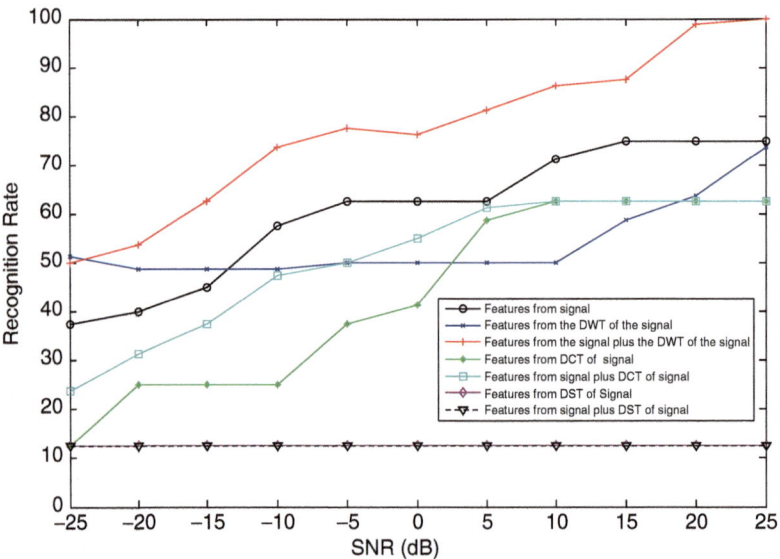

Fig. 121 Recognition rate vs. SNR for the different feature extraction methods with segment-by-segment SVD speech watermarking, speech encryption, and LMMSE deconvolution

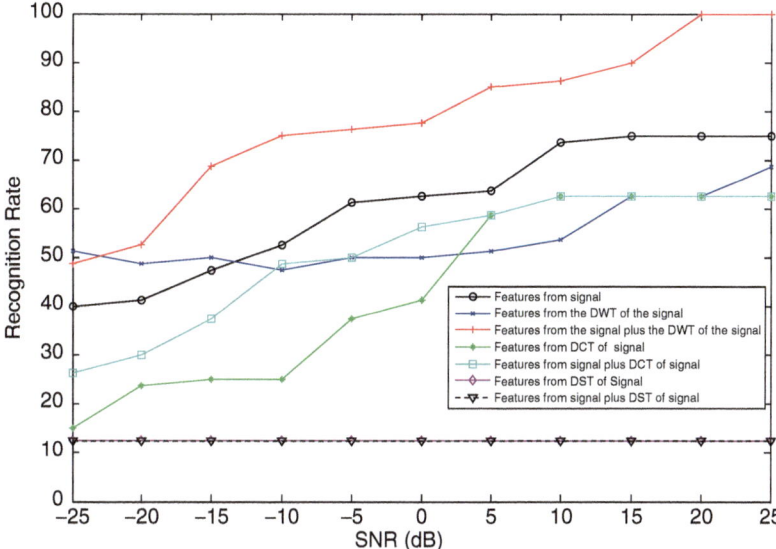

Fig. 122 Recognition rate vs. SNR for the different feature extraction methods with segment-by-segment SVD speech watermarking, speech encryption, and regularized deconvolution

speech deconvolution, and blind signal separation algorithms for performance enhancement of speaker identification systems. A new framework for speaker identification with multilevels of security was also presented in this book. This framework can be developed for real speech signature systems.

11 Directions for Future Research

Future research can be developed in the following directions:

1. Study of speech processing and security techniques for speaker identification systems implementing hidden Markov models, support vector machines, or Gaussian mixture models.
2. Development of sophisticated enhancement, deconvolution, or signal separation algorithms to achieve better enhancement in speaker identification systems performance.
3. Comparison study between speech watermarking and encryption schemes that can be recommended for speaker identification systems with multilevels of security.

References

1. J. P. Campbell, "Speaker Recognition: A tutorial," Proceedings of the IEEE, Vol. 85, No. 9, pp. 1437–1462, 1997.
2. D. A. Reynolds, "An overview of Automatic Speaker Recognition Technology," Proceedings IEEE international conference on Acoustics, Speech and Signal Processing (ICASSP), Vol. 4, pp. 4072–4075, May 2002.
3. G. Saha, P. Kumar, "A comparative Study of Feature Extraction Algorithms on ANN Based Speaker model for speaker Recognition Application," LNCS, Springer-verlag Berlin, Heidelberg, Vol. 3773, pp. 1192–1197, 2004.
4. J. R. Lara, "A Method of Automatic Speaker Recognition using Cepstral Features and Vectorial Quantization," (LNCS), Spring Berlin Heidelberg, pp. 146–153. 2005.
5. F. G. Hashad, T. M. Halim, S. M. Diab, B. M. Sallam, and F. E. Abd El-Samie, "Fingerprint Recognition Using Mel-Frequency Cepstral Coefficients," Pattern Recognition and Image Analysis, Vol. 20, No. 3, pp. 360–369, 2010, Pleiades Publishing, Ltd.
6. F. E. Abd El-samie, "Detection of Landmines From Acoustic Images Based on Cepstral Coefficients," Journal of Sensing and Imaging, Vol. 10, No. 3–4, pp. 63–77, 2009, Springer.
7. U. S. Khan, W. Al-Nuaimy, F. E. Abd El-Samie, "Detection of landmines and underground utilities from acoustic and GPR images with a cepstral approach," Journal of Visual Communications and Image Representation, Vol. 21, No. 7, pp. 731–740, 2010, Elsevier.
8. H. Kasban, O. Zahran, H. Arafa, M. El-Kordy, S. M. S. Elaraby, F. E. Abd El-Samie, "Welding defect detection from radiography images with a cepstral approach," NDT & E International, Vol. 44, No. 2, pp. 226–231, 2011, Elsevier.
9. R. R. Elsharkawy, S. El-Rabaie, M. Hindi, M. I. Dessouky, R. S. Ghoname, "FET Small-Signal Modelling Based on The DST And Mel Frequency Cepstral Coefficients," Progress In Electromagnetics Research B, Vol. 18, pp. 185–204, 2009.
10. R. R. Elsharkawy, S. El-Rabaie, M. Hindi, M. I. Dessouky, R. S. Ghoname, "Neural FET Small-Signal Modelling Based on Mel-Frequency Cepstral Coefficients," Proceedings of the International Conference on Computer Engineering & Systems (ICCES), Cairo, Egypt, pp. 321–326, 2009.
11. R. R. Elsharkawy, S. El-Rabaie, M. Hindi, M. I. Dessouky, "FET Small-Signal Modeling Using Mel-Frequency Cepstral Coefficients and The Discrete Cosine Transform," Journal of Circuits, Systems and Computers, Vol. 19, No. 8 (2010) 1835–1846 World Scientific.
12. R. R. Elsharkawy, S. El-Rabaie, M. Hindi, M. I. Dessouky, "A New Approach For FET Small Signal Modelling Based on Cepstral Coefficients and Discrete Transforms," International Journal of Electronics, Vol. 98, No. 3, March 2011, pp. 379–391, Taylor and Francis.
13. A. Shafik, S. M. Elhalafawy, S. M. Diab, B. M. Sallam and F. E. Abd El-samie, "A Wavelet Based Approach for Speaker Identification from Degraded Speech," International Journal of Communication Networks and Information Security (IJCNIS), Vol. 1, No. 3, pp. 53–60, 2009.
14. M. Hossain, B. Ahmed, M. Asrafi, "A real Time Speaker Identification using artificial Network," 10th international conference on Computer and Information Technology 2007, ICCIT2008, pp. 1–5, Dec. 2007, Dhaka.
15. T. Kinnunen, "Spectral Features For Automatic Test Independent Speaker Recognition," Licentiate's Thesis, University of Joensuu, Department of Computer Science, Finland, 2003.
16. K. L. Neville, Z. M. Hussain, "Effect of Wavelet Compression of Speech on its Mel Cepstral Coefficients", international Conference on Communication, Computer and Power (ICCCP'09), Muscat, Feb. 2009.
17. R. Vergin, D. O'Shaughnessy, A. Farhat, "Generalized Mel- Frequency Cepstral Coefficients for Large-Vocabulary Speaker-Independent Continuous-Speech Recognition" IEEE Transaction on Speech and Audio Processing, Vol. 7, No. 5, pp. 525–532, 1999.
18. H. Wei, C.F. Cheong, C.S. Chiu, "An Efficient MFCC Extraction Method in Speech Recognition," Proceedings of the IEEE International Symposium on Circuits and Systems, ISCAS2006, pp. 145–148, May 2006, Island of Kos, Greece.

19. D. G. Childers, D. P. Skinner, "The Cepstrum: A guide to processing", Proceedings of IEEE, Vol. 65, No. 10, pp. 1428–1443, 1999.
20. T. Matsui and S. Furui, "Comparison of Text-Independent Speaker Recognition Methods Using VQ-Distortion and Discrete/Continuous HMMs," IEEE Transactions on Speech and Audio Processing, Vol. 2, No. 3, pp. 456–459, July 1994.
21. S. Furui, "Cepstral Analysis Technique for Automatic Speaker Verification," IEEE Transactions of Acoustics, Speech, and Signal Processing, Vol. ASSP-29, No. 2, pp. 254–272, 1981.
22. A. I. Galushkin, Neural Networks Theory, Springer-Verlag, Berlin Heidelberg, 2007.
23. G. Dreyfus, "Neural Networks Methodology and Applications," Springer-Verlag, Berlin Heidelberg, 2005.
24. S. L. Bernadin, S. Y. Foo, "Wavelet Processing for Pitch Period Estimation," Proceedings of the 38th Southeastern Symposium on System Theory Tennessee Technological University Cookeville, TN, USA, pp. 5–7, March 2006.
25. V. Krishnan, A. Jayakumar, "Speech Recognition of Isolated Malayalam Words Using Wavelet Features and Artificial Neural Network," 4th IEEE International Symposium on Electronic Design, Test & Applications, DELTA2008, pp. 240–243, 2008, Hong Hong.
26. Z. Tufekci and J. N. Gowdy, "Feature Extraction Using Discrete Wavelet Transform for Speech Recognition", Proceedings of the IEEE, Southeastcon 2000, pp. 116–123, April 2000, Nashville, TN, USA.
27. S. Malik, F. A. Afsar, "Wavelet Transform Based Automatic Speaker Recognition", Proceedings of the IEEE 13th International Multitopic Conference (INMIC2009), pp. 1–4, 2009, Islamabad.
28. I. Daubechies, "Where Do Wavelets Come From?—A Personal Point of View," Proceedings of the IEEE, Vol. 84, No. 4, pp. 510–513, 1996.
29. A. Pizurica, "Image Denoising Using Wavelets and Spatial Context Modeling," Ph. D. Thesis at Ghent University, Belgium, 2002.
30. A. Cohen and J. Kovacevec, "Wavelets: The Mathematical Background," Proceedings of the IEEE, Vol. 84, No. 4, pp. 514–522, 1996.
31. N. H. Nielsen and M. V. Wickerhauser, "Wavelets and Time-Frequency Analysis," Proceedings of the IEEE, Vol. 84, No. 4, pp. 523–540, 1996.
32. K. Ramchndran, M. Vetterli and C. Herley, "Wavelets, Subband Coding, and Best Basis," Proceedings of the IEEE, Vol. 84, No. 4, pp. 541–560, 1996.
33. M. Unser and A. Aldroubi, "A Review of Wavelets in Biomedical Applications," Proceedings of the IEEE, Vol. 84, No. 4, pp. 626–638, 1996.
34. W. Sweldens, "Wavelets: What Next?," Proceedings of the IEEE, Vol. 84, No. 4, pp. 680–685, 1996.
35. A. Prochazka, J. Uhlir, P. J. W. Rayner and N. J. Kingsbury, Signal Analysis and Prediction. Birkhauser Inc., New York, 1998.
36. J. S. Walker, A Primer on Wavelets and Their Scientific Applications. CRC Press LLC, Boca Raton, 1999.
37. A. Prochazka, J. Uhlir, P. J. W. Rayner and N. J. Kingsbury, Signal Analysis and Prediction. Birkhauser Inc., New York, 1998.
38. M. Farge, N. Kevlahan, V. Perrier and E. Goirand, "Wavelets and Turbulence," Proceedings of the IEEE, Vol. 84, No. 4, pp. 639–669, 1996.
39. R. Kubichek, "Mel-Cepstral Distance Measure for Objective Speech Quality Assessment," Proceedings of The IEEE Pacific Rim Conference on Communications, Computers and Signal Processing, pp. 125–128, 1993, Victoria, BC.
40. S. Wang, A. Sekey, A. Gersho, "An Objective Measure for Predicting Subjective Quality of Speech Coders," IEEE Journal on selected areas in communications, Vol. 10, No. 5, pp. 819–829, 1992.
41. W. Yang, M. Benbouchta, R. Yantorno, "Performance of the Modified bark Spectral Distortion as an Objective Speech Quality Measure," Proceedings of the IEEE International Conf. on Acoustic, Speech and Signal Processing (ICASSP), Vol. 1, Washington, USA, pp. 541–544. 1998.

42. R. E. Crochiere, J. E. Tribolet, and L. R. Rabiner, "An Interpretation of The Log Likelihood Ratio as a Measure of Waveform Coder Performance," IEEE Transactions on Acoustics, Speech, And Signal Processing, Vol. Assp-28, No. 3, pp. 318–323, 1980.

43. N. W. D. Evans, J. S. D. Mason, "An Assessment on the Fundamental Limitations of Spectral Subtraction", IEEE international conference on Acoustic, speech and signal processing, Vol. 1, pp. 1–1, 2006.

44. S. F. Boll, "Suppression of Acoustic Noise in Speech using Spectral Subtraction", IEEE Trans. ASSP, Vol. 27 (2), pp. 113–120, 1979.

45. P. Krishnamoorthy, S. R. Mahadeva, "Enhancement of Noisy Speech by Spectral Subtraction and Residual Modification", Proceedings of the IEEE Annual India conference, pp. 1–5, 2006, New Delhi.

46. M. A. Abd El-Fattah, M. I. Dessouky, S. M. Diab and F. E. Abd El-Samie, "Speech Enhancement Using An Adaptive Wiener Filtering Approach," Progress In Electromagnetics Research M, Vol. 4, 167–184, 2008.

47. S. E. El-Khamy, M. M. Hadhoud, M. I. Dessouky, B. M. Salam, and F. E. Abd El-Samie, "Regularized Super-Resolution Reconstruction of Images Using Wavelet Fusion," Journal of Optical Engineering, Vol. 44, No. 9, 2005, SPIE.

48. S. E. El-Khamy, M. M. Hadhoud, M. I. Dessouky, B. M. Salam, and F. E. Abd El-Samie, "Wavelet Fusion: A Tool to Break The Limits on LMMSE Image Super-Resolution," International Journal of Wavelets, Multiresolution and Information Processing, Vol. 4, No. 1, 2006, pp. 105–118, World-Scientific.

49. S. E. El-Khamy, M. M. Hadhoud, M. I. Dessouky, B. M. Salam, and F. E. Abd El-Samie, "A Wavelet Based Entropic Approach To High Resolution Image Reconstruction," International Journal of Machine Graphics & Vision, Vol. 17, No. 4, pp. 235–256, 2008.

50. D. C. Chan, "Blind Signal Separation" A PhD dissertation, University of Cambridge, January 1997.

51. H. Hammam, A. Abou Elazm, M. E. Elhalawany, F. E. Abd El-Samie, "Blind separation of audio signals using trigonometric transforms and wavelet denoising," International Journal of Speech Technology, Vol. 13, No. 1, pp. 1–12, 2010, Springer.

52. S. E. El-Khamy, M. M. Hadhoud, M. I. Dessouky, B. M. Sallam, and F. E. Abd El-Samie, "Enhanced Wiener Restoration of Images Based on The Haar Wavelet Transform," Proceedings of the URSI National Radio Science Conference (NRSC), Manssoura, Egypt, March 2001.

53. S. E. El-Khamy, M. M. Hadhoud, M. I. Dessouky, and F. E. Abd El-Samie, "A New Technique For Enhanced Regularized Image Restoration," Proceedings of the URSI National Radio Science Conference (NRSC), Alexandria, Egypt, March 2002.

54. S. E. El-Khamy, M. M. Hadhoud, M. I. Dessouky, B. M. Salam and F. E. Abd El-Samie, "Sectioned Implementation of Regularized Image Interpolation" Proceedings of the 46th Proceedings of IEEE MWSCAS, Cairo, Egypt, Dec. 2003.

55. S. E. El-Khamy, M. M. Hadhoud, M. I. Dessouky, B. M. Salam, and F. E. Abd El-Samie, "Optimization of Image Interpolation as an Inverse Problem Using The LMMSE Algorithm," Proceedings of the IEEE MELECON, pp. 247–250, Croatia, May 2004.

56. M. M. Hadhoud, F. E. Abd El-Samie and S. E. El-Khamy, "New Trends in High Resolution Image Processing" Proceedings of the IEEE Fourth Workshop on Photonics and Its Application, National Institute of Laser Enhanced Science, Cairo University, Egypt, May 2004.

57. A. K. Jain, "Fast Inversion of Banded Toeplitz Matrices by Circular Decomposition" IEEE Trans. Acoustics, Speech and Signal Processing, Vol. ASSP-26, No. 2, pp. 121–126, April 1978.

58. H. C. Anderws and B. R. Hunt, Digital Image Restoration. Englewood Cliffs, NJ: Prentice-Hall, 1977.

59. B. Macq, J. Dittmann, and E. J. Delp, "Benchmarking of Image Watermarking Algorithms for Digital Rights Management" Proceedings of The IEEE, Vol. 92, No. 6, pp. 971–984, 2004.

60. Z. M. Lu, D. G. Xu, and S. H. Sun, "Multipurpose Image Watermarking Algorithm Based on Multistage Vector Quantization," IEEE Transactions on Image Processing, Vol. 14, No. 6, pp. 822–831, 2005.
61. H. S. Kim and H. K. Lee, "Invariant Image Watermark Using Zernike Moments," IEEE Transactions on Circuits And Systems For Video Technology, Vol. 13, No. 8, pp. 766–775, 2003.
62. W. C. Chu, "DCT-Based Image Watermarking Using Subsampling," IEEE Transactions on Multimedia, Vol. 5, No. 1, pp. 34–38, 2003.
63. L. Ghouti, A. Bouridane, M. K. Ibrahim, and S. Boussakta, "Digital Image Watermarking Using Balanced Multiwavelets," IEEE Transactions on Signal Processing, Vol. 54, No. 4, pp. 1519–1536, 2006.
64. S. Xiang and J. Huang, "Histogram-Based Audio Watermarking Against Time-Scale Modification and Cropping Attacks," IEEE Transactions on Multimedia, Vol. 9, No. 7, pp. 1357–1372, 2007.
65. Z. Liu and A. Inoue, "Audio Watermarking Techniques Using Sinusoidal Patterns Based on Pseudorandom Sequences," IEEE Transactions On Circuits And Systems For Video Technology, Vol. 13, No. 8, pp. 801–812, 2003.
66. A. N. Lemma, J. Aprea, W. Oomen, and L. V. de Kerkhof, "A Temporal Domain Audio Watermarking Technique," IEEE Transactions on Signal Processing, Vol. 51, No. 4, pp. 1088–1097, 2003.
67. W. Li, X. Xue, and P. Lu, "Localized Audio Watermarking Technique Robust Against Time-Scale Modification," IEEE Transactions on Multimedia, Vol. 8, No. 1, pp. 60–69, 2006.
68. S. Erküçük, S. Krishnan, and M. Z. Glu, "A Robust Audio Watermark Representation Based on Linear Chirps," IEEE Transactions on Multimedia, Vol. 8, No. 5, pp. 925–936, 2006.
69. X. Wang, W. Qi, and P. Niu, "A New Adaptive Digital Audio Watermarking Based on Support Vector Regression" IEEE Transactions on Audio, Speech, And Language Processing, Vol. 15, No. 8, pp. 2270–2277, 2007.
70. F. E. Abd El-samie, "An efficient singular value decomposition algorithm for digital audio watermarking," International Journal of Speech Technology, Vol. 12, No. 1, pp. 27–45, 2009, Springer.
71. W. Al-Nuaimy, M. A. M. El-Bendary, A. Shafik, F. Shawki, A. E. Abou El-azm, N. A. El-Fishawy, S. M. Elhalafawy, S. M. Diab, B. M. Sallam, F. E. Abd El-Samie, "An SVD Audio Watermarking Approach Using Chaotic Encrypted Images," Accepted for publication in Digital Signal Processing, Elsevier, doi:10.1016/j.dsp.2011.01.013.
72. I. F. Elashry, O. S. Farag Allah, A. M. Abbas, S. El-Rabaie and F. E. Abd El-Samie, "Homomorphic Image Encryption," Journal of Electronic Imaging, Vol. 18, No. 3, 033002, 2009, SPIE.
73. R. Liu and T. Tan, "An SVD-Based Watermarking Scheme for protecting rightful ownership", IEEE Trans. on Multimedia, Vol. 4, No. 1, March, 2002, 121–128.
74. E. Ganic and A. M. Eskicioglu, "Secure DWT-SVD Domain Image Watermarking: Embedding Data in All Frequencies," ACM Multimedia and Security Workshop, Magdeburg, Germany, September 20–21, 2004.
75. R. A. Ghazy, N. A. El-Fishawy, M. M. Hadhoud, M. I. Dessouky and F. E. Abd El-Samie, "An Efficient Block-by-Block SVD-Based Image Watermarking Scheme," Proceedings of the URSI National Radio Science Conference, NRSC, Cairo, Egypt, March 2007.
76. J. M. Shieh, D. C. Lou, and M. C. Chang, "A Semi-Blind Watermarking Scheme Based on Singular Value Decomposition," Computer Standards & Interface, 28, 428–440, 2006.
77. F. Han, X. Yu, and S. Han, "Improved Baker Map for Image Encryption" Proceedings of the First International Symposium on Systems and Control In Aerospace and Astronautics, ISSCAA 2006, Harbin, pp. 1273–1276.
78. J. Fridrich, "Image Encryption Based On Chaotic Maps," Proceedings of the IEEE International Conference on Systems, Man, and Cybernetics, Orlando, 1997, pp. 1105–1110.

79. Q. Qian, Z. Chen, Z. Yuan, "Video Compression and Encryption Based on Multiple Chaotic System," Proceedings of The IEEE 3rd International Conference on Innovative Computing Information and Control (ICICIC'08), Dalian, Liaoning.

80. F. Huang, F. Lei, "A Novel Symmetric Image Encryption Approach Based on a New Invertible Two-dimensional Map," Proceedings of the International Conference on Intelligent Information Hiding and Multimedia Signal Processing IIHMSP 2008, pp. 1340–1343, Harbin.

81. S. C. Koduru and V. Chandrasekaran, "Integrated Confusion-Diffusion Mechanisms for Chaos Based Image Encryption," Proceedings of the IEEE 8th International Conference on Computer and Information Technology Workshops, 2008, pp. 260–263, Sydney.

82. K. Usman, H. Juzojil and I. Nakajimal, "Medical Image Encryption Based on Pixel Arrangement and Random Permutation for Transmission Security," Proceedings of the IEEE 9th International Conference on e-Health Networking, Application and Services, 2007, pp. 244–247, Taipei.

83. E. S. Hassan, X. Zhu, S. E. El-Khamy, M. I. Dessouky, S. A. El-Dolil, F. E. Abd El-Samie, "A Chaotic Interleaving Scheme for the Continuous Phase Modulation Based Single-Carrier Frequency-Domain Equalization System," Accepted For Publication in Wireless Personal Communications, Springer, DOI 10.1007/s11277-010-0047-z.

84. J. F. Andrade, M. L. Campos and J. A. Apolinario, "Speech Privacy for Modern Mobile Communication Systems," IEEE International Conference on Acoustics, Speech and Signal Processing, 2008, Las Vegas, NV.

85. B. Goldburg, S. Sridharan, and E. Dawson, "Design and Cryptanalysis of Transform Based Analog Speech Scramblers," IEEE Journal of Selected Areas on Communications, Vol. 11, pp. 735–743, 1993.

86. E. Mosa, N. W. Messiha, O. Zahran, F. E. Abd El-Samie, "Encryption of speech signal with multiple secret keys in time and transform domains," International Journal of Speech Technology, Vol. 13, No. 4, pp. 231–242, 2010, Springer.

87. N. El-Fishawy and O. M. Abu Zaid, "Quality of Encryption Measurement of Bitmap Images with RC6, MRC6, and Rijndael Block Cipher Algorithms," Int. J. Network Security, Vol. 5, No. 3, pp. 241–251, 2007.

88. C. E. Shannon, "Communication Theory of Secrecy System", Bell Syst. Tech. J, Vol. 28, pp. 656–715, 1949.

89. C. J. Kuo, "Novel Image Encryption Technique and Its Application in Progressive Transmission," J. Electronic Imaging, Vol. 2, No. 4, pp. 345–351, 1993.